Water and Scriptures

K.V. Raju • S. Manasi

Water and Scriptures

Ancient Roots for Sustainable Development

Springer

Editors
K.V. Raju
Policy and Impact
Asia Regional Program
International Crops Research Institute
for the Semi-Arid Tropics
Hyderabad, Telangana, India

S. Manasi
Centre for Research in Urban Affairs
Institute for Social and Economic
Change (ISEC)
Bengaluru, Karnataka, India

ISBN 978-3-319-50561-9 ISBN 978-3-319-50562-6 (eBook)
DOI 10.1007/978-3-319-50562-6

Library of Congress Control Number: 2017932533

Printed on acid-free paper

This Springer imprint is published by Springer Nature
The registered company is Springer International Publishing AG
The registered company address is: Gewerbestrasse 11, 6330 Cham, Switzerland

Preface

Cultures change over time all over the world; however, it is important to see how we adapt to cultural changes while meeting the needs of our society without affecting environmental sustenance. Several experts have emphasized the role of culture in protecting the environment. We mulled over this thought and believed that consolidation of traditional knowledge and its understanding will go a long way in improving the current environmental challenges besides aiding in environmental rehabilitation, conservation and improved resource management. Culture and religion are often interlinked. Also, water and culture are strongly interlinked. Water is a vital source of life and culture determines a person's way of life. Keeping this in perspective, since ancient scriptures often highlight the importance of a harmonious coexistence between human beings and nature, we have attempted to draw lessons from the scriptures of three religions, Islam, Christianity and Hinduism, for water in specific. We plan to extend this framework to other prominent religions like Judaism, Buddhism, Zoroastrianism and Shinto, and also to other natural resources in future years.

There are several studies related to scriptures across religions and their perceptions regarding ecological conservation. However, religious studies and their socio-economic and environmental relevance to society, more specifically to the current policy contexts, are limited. This study attempts to bridge this gap. Given this setting, the focus was to explore how water was conceptualized, conserved and managed in scriptures, and to understand the historical, functional and futuristic perspectives of water resources management to develop policy guidelines.

The book will be useful for historians and research scholars studying water sharing and the place of water in different cultures as well as ecologists and environmental scientists. The chapters provide messages from religious scriptures and their relevance for sustainable water resources management in Islam, Christianity and Hinduism. Legal perspectives are spread across all the scripture-based chapters, followed by a comparative overview across religious scriptures and policy perspectives. We involved scholars from respective religions well versed in the ancient scriptures in writing this book.

We are thankful to the Institute for Social and Economic Change, Bengaluru, for giving us the opportunity to carry out the study. Also, we are extremely thankful to all authors, reviewers, scripture experts, and colleagues who have extended support and contributions all throughout the study. Several deliberations with subject experts and scholars were held during the consultative workshops held at the research and study centres of Islamic, Christian and Hindu religions and a seminar was held before finalization. We remain indebted to all of them.

Hyderabad, TS, India K.V. Raju
Bengaluru, KA, India S. Manasi

Contents

Contents

Water and Scriptures: An Introduction

K.V. Raju and S. Manasi

Introduction

Across the world, various cultures and their practices have influenced, for thousands of years, the way people perceive and practice the use and abuse of natural resources. Relations between people and their environments are embedded in culture (Schelwald-van der Kley and Reijerkert 2009). 'Cultures' change over time all over the world; however, it is important to see how we adapt to cultural changes while meeting the needs of our society without affecting the environmental sustenance. While an emphasis on a wide variety of ethical principles is seen as a means to promote the cause of environment, the acceptable options available to people depend on the cultural contexts of the societies concerned. Ethical concerns are seen as part of a cultural evolution of people, including environmental ethics (Nadkarni 2011). For instance, the ancient scriptures often emphasize the importance of a harmonious coexistence between human beings and nature. '*Do strengthen me. May all beings regard me with eyes of a friend. May I regard all beings with the eyes of a friend. With the eyes of a friend do we regard one another*'. YV 36.18 (Sharma 2000). Moses (refer Chapter 2.2.3), drawing from the Bible, indicates that in the Bible, the word of God and water are considered important for life, compares the word of God with 'rain'. The believer is called to accept the word and water, cherish, respect and honour them both as the gifts of God upon freely bestowed. Water becomes a symbol for the outpouring of God's spirit and blessing. God "will pour water on the thirsty land and streams on the dry ground," *Isaiah 44, 3*. Similarly, Siraj and Tayab

K.V. Raju (✉)
Policy and Impact, Asia Regional Program, International Crops Research Institute for the Semi-Arid Tropics, Hyderabad, India

S. Manasi
Centre for Research in Urban Affairs, Institute for Social and Economic Change (ISEC), Bengaluru, India
e-mail: manasi@isec.ac.in

K.V. Raju, S. Manasi (eds.), *Water and Scriptures*,
DOI 10.1007/978-3-319-50562-6_1

(refer Chapter 2.1) draws from the Quran that asks us not to waste (water, food, etc.). It states … 'But waste not by excess for God loves not the wasters' 6:141. All this signifies the cultural significance and as rightly pointed out by Alveré (2012) highlighting different key characteristics of culture and explains how culture manifests itself at different stages in life, affects the behavioural pattern of people and its interpretation and influences biological processes as well. Currently, natural resources management is planned in isolation resulting in piecemeal approaches. Cultural aspects that play an important role affecting behavioural patterns are not included, thus making sustainable management of resources a futile exercise in several situations.

In the recent decades, the climate change challenge encountered is a serious concern across the globe. Several national and international efforts are intended towards combating the effects of climate change. Climate variations have always been witnessed with natural influences, however, lately; there are strong evidences that human actions are main drivers of the recent increase in global temperatures that affect precipitation patterns and extreme weather events. The fifth Intergovernmental Panel on Climate Change (IPCC) report states with 95% confidence that humans are the main cause of the current global warming. According to NASA and NOAA data, 2015 was the hottest year and since 1992, almost every year has been indicating an increasing trend. Water, agriculture, energy and climate change are all related, and their harmony is important for sustainable development. With growing disharmony across these sectors, earth will be facing grave consequences.

The present threats of climate change are widely discussed in the context of bringing about a collective behavioural change. Based on American Psychological Association's report on Climate Change, an interface between Psychology and Global Climate Change points out ways psychology can aid people adopt more environmentally friendly behaviour and ways to incorporate it into policy and behavioural research to strengthen environmental laws and regulations (Price 2009). There is a broad consensus that global warming is caused due to anthropogenic factors and will continue if greenhouse gas emissions are not radically reduced. Several initiatives in resource management, technological interventions, economic initiatives, law and regulations, financing environmental challenges and several other approaches have not been able to completely resolve the crisis. This brings us to the stage wherein there is need for more than just business-as-usual solutions mentioned above.

All this brings to the fore the need to explore a holistic stewardship approach to seeking effective solutions to mitigate environmental challenges. It would be useful to understand the key changes in existing value systems, cultural practices and mindsets. There is still lack of proper understanding about the causes of anthropogenic factors on environment from the cultural perspectives on the problems that are prevailing in current contexts and what religion has got to offer to bring about the needed change. Based on the learnings from scriptures, we intend to evolve policy options for basing natural resources management that is inclusive of cultural understanding and contexts so that cultural techniques may be included in addressing current challenges. To begin with, we will focus on 'water' the critical element for sustainer of life since it is one of the resources that is under serious threat paving the world towards severe water crisis. According to the United Nations (2006), water use has grown more than twice the rate of population increase in the last century. By 2025, an estimated 1.8 billion people will live in areas plagued by water scarcity, with two-third of the world's population living in water-stressed regions resulting from use, growth and climate change.

Peter Gleick in his article State of the Planet says, 'The most serious unresolved water problem is the continued failure to meet the basic human need for water'. Thus, water resources management demands a holistic approach across the globe, given the fact that water knows no boundaries. Water has several functions—economic, ecological, societal and spiritual. Hence, water management needs to be sustainable. Schelwald-van der Kley and Reijerkert (2009) argue that water plays a fundamental role in the life of societies; it has a strong cultural dimension and no sustainable solutions can be found by omitting this dimension. They stress that intimate relationship between water and people should be considered during decision-making processes. The way water is conceived, valued, understood, managed, used or abused, worshipped or desecrated are all influenced by the cultures. We too see the immense scope for integrating the cultural dimension, however; we will focus specifically to 'water' in 'religious scriptures'. We see that wise management of water based on the spiritual principles of justice and equity integrating other disciplines and stakeholders would play an important role in sustainable water use.

As per the UNESCO report (2003), *'Water is probably the only natural resource to touch all aspects of human civilization—from agricultural and industrial development to the cultural and religious values embedded in society. The need and demand for water have been a driving force for health, for society, for economic prosperity, for cultural significance and development throughout human history'.* Cultural differences play a key role in the way water is perceived, valued and managed in different societies. All views on water are linked to its fundamental functions and its relevance to life. The fact that water is necessary for survival makes it a key part of culture. Moreover, all major civilizations developed around large sources of water and mostly around sources of freshwater.

The significance of water and scriptures in a nutshell describes water as a physical substance, a biological necessity, an intrinsic part of people's identities, cultures, religious perceptions of themselves and the Otherworld or the life thereafter. Water in its many facets matters for humans, while the social, cultural, ideological and religious roles of water include deep ontological relations and identities ranging from personal perceptions and gender relations to rainmaking and fertility rites for the benefit of the whole society as well as perceptions of cosmological realms and religious beliefs (Oestigaard 2009).

Drawing from the messages across the World Water Day conferences held by the UNESCO, several dimensions depicting the inter-linkages between water and culture were highlighted. It emphasizes the relevance of culture in protecting water resources. The universal declaration of the 2002 Paris World Water Day stresses that water and culture are intertwined, hence, *'Promote understanding that Water and Culture are inseparable elements of human life. Culture should be regarded as the permanently evolving set of distinctive spiritual, material, intellectual and emo-*

tional features of society or a social group. It encompasses—in addition to art and literature, lifestyles, ways of living together, value systems, traditions and beliefs (UNESCO Universal Declaration on Cultural Diversity, UNESCO, Paris 2002)', as a follow-up of this, a key message at the World Water Forum (2003) held at Japan, highlighted water's cultural importance for sustainable management thus, '*Due to its fundamental role in society's life, water has a strong cultural dimension. Without understanding and considering the cultural aspects of our water problems, no sustainable solution can be found'—Session—Water and Cultural Diversity, Statement to the Ministerial Conference, 3rd World Water Forum, 22 March 2003.* The World Water Day (WWD) (2006) also focused on the theme 'Water and Culture' under the leadership of the UNESCO. Several messages brought out in a summation focused on '*promoting inclusive and solution oriented water governance and encouraging artistic expressions on water issues for creating awareness, understanding and sharing information'*.

Another interesting initiative relates to the recommendation made about the need for a '*Water Ethics Charter*' at the 2012 World Water Forum in Marseille in the session, 'Towards Ethical and Spiritual Approach to Water'. Several organizations worked in collaboration to develop a declaration prescribing a set of ethical principles to guide decisions on water use and water ecosystem management. The Charter was presented at the World Water Forum in South Korea in March 2015.

Water, Culture and Religion

Culture and religion are often interlinked. Also, water and culture are also strongly interlinked. Water is a vital source of life and culture determines a person's 'way of life'. This is inclusive of the way people manage their water resources, given the environment they live in and the way they adapt to it. Indigenous cultures are known even to this day for their sustainable water practices. For instance, in many traditional cultures, water is linked to people's identity, if anyone destroys their water resource it amounts to depriving them of their cultural identity. Similarly, in few other cultures, water has an important social function, particularly for women. Most often, modern practices have disturbed these traditional practices resulting in negative implications. However in some cases, there has been revival of old traditions for natural and sustainable water use. For instance, traditional Karez and Qanat systems found in arid countries like Afghanistan and Pakistan are currently being restored to use. These systems are owned and operated by the local community where the irrigation channels built underground brings water from the mountains to the villages, preventing evaporation losses. Thus, sustainable water management has been a practice that cuts across cultures where there prevails a delicate balance between water resources and human society (Schelwald-van der Kley and Reijerkert 2009).

From the religions front, Elaide argues that religious phenomena will only be recognized as such if it is grasped at its own level, hence, to be studied as something religious. This is due to the fact that grasping the essence of such a phenomenon through physiology, psychology, sociology, economics, linguistics, art or any other study would remain untrue as it misses the most unique element—'sacred'. Obviously, there are no purely religious phenomena, so phenomena can be solely and extremely religious because religion is human; she discusses by studying an account of certain cosmic hierophanies like the 'sacred' revealed at different cosmic levels—sky, water, earth, stones. For instance, water and the seeds of things, water symbolizes the whole of potentiality. In Sumerian, water means 'sperm, conception, generation', in Mesopotamian carvings, the symbol fish and water are seen as emblems of fertility, while in China, the dragon—an emblem of sky and water was constantly associated with the emperor, who represented the rhythms of the cosmos and conferred fecundity on the earth (Elaide 1996).

Besides this, the ancient scriptures across religions have related literature that have described the nature of these resources and guiding people for centuries, i.e. the way people should treat these resources which are part of their daily life and rituals being practised. As observed by Nadkarni (2011), 'reverence for nature was not confined to preaching in the scriptures and texts, but was part and parcel of their traditional culture'. The Vedas, Upanishads, Puranas, Epics and scholarly writings such as 'Mayurchitraka' and 'Brihat Samhita' are vast treasure troves of scientific and environmental knowledge. The ancient scriptures, particularly the Vedic literature stresses maintaining the importance of natural resources. Protection of environment is articulated through informal rules in the form of traditions/daily rituals/prayers along with compassion towards all forms of life with a particular emphasis on human well-being wherein pious actions would result in nature's blessings, (the earth—*mother* and the atmosphere above—*father* would protect and provide one with a long and peaceful life) while erroneous actions could harm nature and its functions (Nair 2003). Water is considered an inspiration since centuries and people have deep rooted spiritual and religious values and strong faith that unite them and sustain them. Cultural views on water resource management are largely based on the main religious views of the region. Water holds a fundamental place in most religious beliefs, values and rituals. Indigenous spiritual and religious beliefs hold a profound reverence for water, and the customary way of life does offer an exemplar for a sustainable way of life. Therefore, dialogues with reference to water resources management in religion become applicable for discussions relating to water resources management in culture.

Among these, water plays an important role not only in five major religions of the world, but also almost every recognized religion in history. Animism, Judaism, Taoism, Baha'is, religions have all emphasized on the respect to water. In Animism, indigenous people have honoured and respected water as sacred, hence, the traditional knowledge, their laws have made them feel responsible for protecting water as it connects all forms of life. In Buddhism, water symbolizes purity, clarity and calmness. Judaism considers water as an important component for ritual cleansing practices. Taoism sees water as the essence of nature and a model for human conduct. Baha'is religion believes that water management as the notion of unity between all people and the interconnectedness of things. Similarly, Hinduism considers water as divine besides a cleanser of sins, Christianity sees water as purifier and much more, while Islam also sees water as life and a gift from God (Schelwald-van der Kley and Reijerkert 2009).

The rituals and beliefs across religions vary, but there are consistent views on water with regard to the foundation of most religions and the natural significance of water. Most religions are based on a set of morals about 'what is right and what is wrong' and most of them view doing wrong things as being 'impure'. As water cleans and purifies, many religions have beliefs and rituals that associate water with the power of cleaning us of our sins or moral impurities (Stewart 2008). 'Rebirth' and 'Purification' are the two main symbolic qualities of water that have made it sacred in the religious ceremonies. In rituals, water confers 'new birth' in magic rituals it heals and funeral rites it assures rebirth after death. Because it incorporates in itself all potentiality, water becomes a symbol of life (living water) rich in seeds it fertilizes (Schelwald-van der Kley and Reijerkert 2009).

Therefore, perceptions about religious and non-religious beliefs of water are critical to proficiently deal with water management issues. Religious wisdom would thus impart wise water management by connecting values and behaviour that would uphold sustainable water management. It would be useful to find the precise blend between the old and modern practices that would aid in sustainable solutions to cope with the current water challenges.

Background and Process

We have been working on this theme since 2011,[1] and this study was taken up to further strengthen the work with the support from scholars well versed in the ancient scriptures. We felt a consolidation of traditional knowledge and its understanding will go a long way in improving the current environmental challenges besides aiding in environmental rehabilitation, conservation, and improved resource management. Examining social value systems and the changing water environments and their impacts on traditions and cultural values are important to comprehending cultural changes in a historical perspective. It is important that we include culture-based learnings for developing cultural models as an important component to achieving sustainable development. Innovative and specific technological interventions have also evolved as a good alternative in the recent times. For instance, in Delhi, 'Temple machine' is used at few temples to make 'holy compost' out of the temple waste generated and is being sold to devotees which is an interesting initiative, thus reducing the waste dumping into the river Yamuna. Ganesha idols made from clay implanted with plant seeds, thus, every Ganesha idol when immersed methodically turns into a plant. Several such examples can be adopted and scaled up.

There are several studies related to scriptures across religions and their perceptions regarding ecological conservation. However, religious studies and their socio-economic and environmental relevance to society, more specifically to the current policy contexts, are limited. Scholarly writings on the subject have, in the past, invariably, glorified the 'golden past' or vilified it as being dogma ridden and unscientific.

[1] In the Center for Ecological Economics and Natural Resources of the Institute for Social and Economic Change, Bengaluru, India.

Contemporary writings have made fleeting references to the contributions of the past to the present prevailing structure, scheme and form of Water Governance. The Indian Law Institute had, in 1990s, published two scholarly works on Water Law and Governance ('Water Rights and Principles of Water Resources Management', 1991 and 'Water Law in India', 1992, both edited by Chhatrapati Singh) and the recent publication on 'Water and the Law' (edited by RamaswamyIyer, 2009). These are, perhaps, the significant works to have devoted some space to reflecting on the Past Practices on Water Management, but only in general and among many other aspects of concern. Efforts in one sweep, in carrying out a detailed analysis of the customs, traditions and practices across religions (besides the legal ordering on Water in the pre-independence era) and their continued relevance to the present context have hardly been attempted. 'Water and Scriptures', in this respect, appears to have made a significant effort in that direction. There is significant effort made in recounting the past practices, frequently referred to in different religion's religious scriptures and cultural orientations, in a structured manner and connecting them to the present system of water management, in India as learning lessons from the past to effectively address the challenges of the present and future. Given this contexts, our specific aim focused on drawing lessons and apply indigenous knowledge, wisdom and cultural traditions to suit the policy context describing the role of individuals, communities and the state to ensure an effective natural resources management, in this context, water. 'Traditional Practices' threw up the question, 'Tradition, in relation to which modern practice?' as different cultures and traditions slipped into modern times at different points of time, in history. It was then proposed that the research would attempt to look into as many earlier traditions as possible to narrow the focus of enquiry, to the extraction of the theoretical foundations and principles of water governance, in a comparative context globally and elaborate on developments in India, till its independence from the alien rule. As regards India, in particular, it was decided on the cut-off period for the study to include the developments during the Mughal period and the influence of the colonial masters—the British—up to Indian independence.

Given this setting, the focus of this study was: (a) to explore how water were conceptualized in scriptures, civilizations and through dynastic rules to conserve and manage water resources; (b) to understand the historical, functional and futuristic perspectives of water resources management; (c) to draw lessons and evolve policy guidelines.

Process

First Phase

We initiated small studies in a phased manner. To begin with, three papers were taken up for obtaining a holistic view of 'water' in particular Heritage view in India:

- Water and Islamic records
- Water and Biblical records

- Water and Hindu Scriptures
- Water and Law (cuts across all three religious scriptures)

We plan to extend this framework to other prominent religions like Judaism, Buddhism, Zoroastrianism and Shinto in future years. We admit that identifying experts and the process of understanding the expectations that we have from them to suit the framework developed has been a challenging task, so far.

The studies taken up provide an overview of water resources management over time besides attempting to understand the values that the ancient texts and scriptures have laid down with regard to water resources management in specific, further detailing out challenges in water resources management in the current context and suggesting future options for improved water resources management. At a later phase, the focus will be on a comparative study of the cultural practices and their relevance to the present society and policy contexts.

Content Framework

A detailed content framework was developed and circulated to the specific paper writers. The content frame was developed based on an earlier paper, "Water—A Heritage View",[2] where we worked at length on water perceived in its various cultural forms and traditions and related aspects, in addition to reviewing the literature and discussions with experts and our current experiences in the arena of water resources management. We would like to highlight in the current study that culture, which is largely routed through our rituals/practices has a great influence—positive and negative on the challenges encountered in the current context; hence, it is important to analyse the cultural dimension, its relevance and implications to bring about the needed change.

Several instances can be adopted even to this day, for instance, the positives of traditional water quality technologies—practically useful and simple—such as using alum for water purification and sand filters for bringing about a significant change in the water quality. Similarly, it is important to analyse the implications that in materials used in making Ganesha and Durga idols hold for contamination of water bodies, thus, highlighting the need for regulation by providing factual information and creating awareness among the populace, of course, keeping in view the religious sentiments involved in celebrating these festivals. Nadkarni (2011) in his article on Culture and Environment, argues that culture influences our attitudes and behaviour and that it can also be used as a resource to promoting environmentally benign attitudes. Being not rigid, culture can evolve and respond to changes that newer problems bring in their wake. In situations where culture is not initially benign towards environment, leaders of thought should play a role in bringing about

[2] Published under the working paper series of the Development Foundation, Bengaluru (Manasi. S and K. V. Raju).

the action as they have insights into the future. Largely, all indigenous people's attitudes are oriented towards sustainable use of natural resources. Besides, industrialization is seen as one of the major causes underlying the promotion of philosophy of private profit maximization as well as freedom for the market, which tend to disrupt the cultural traditions. With knowledge about the cause of these problems, culture and ethics can evolve towards facing new challenges posed to the environment.

The content frame comprises several sections like—reverence to water in ancient texts, benefits from water, conservation and promotion practices, water purification practices, water festivals, technologies in water management, water use and management in ancient Indian civilizations and so on. In the current context, it covers the status and key concerns in water management—use and abuse in view of cultural practices, pollution, water access, inefficient use, water markets and as a futuristic context, the lessons to be drawn for sustainable practices in water management. Besides, we also held several informal meetings to refine the framework content.

The research universe on water and law evolved to closely align with and complement the design commonly worked for the other three papers on Hinduism, Islam and Bible. The layout for the legal enquiry was specifically required to cover during the said period, (a) legal conceptualization and characterization of water; (b) issues of rights, responsibilities, need, entitlements, livelihood, commerce, prioritization of use, etc.; (c) principles of governance—polluter pays, precaution, conservation, sustainable use, public trust, equity, etc. and strategies for augmentation, improvement of quality and protection; (d) administrative architecture, with roles and responsibilities at different levels and layers of governance and among different players; (e) mechanisms for dealing with deviance of prescriptions and penal sanctions; (f) grievance redressal and conflict-resolution mechanisms; (g) devises for relating, integrating and harmonizing water governance with the governance of other aspects of natural resources management (land, forest, etc.); (h) comparison between and evaluation of different legal systems and governance and (i) identification of the good prescriptions and practices of the past and the extent and nature of the survival of such content in the current situation.

Progress Review

We had constituted a review committee consisting[3] who were part of the whole process from the time of initiation providing valuable feedback on the papers and refining the process as well. Discussions were held with the experts in the respective religion. We circulated a draft framework prior to holding the meetings.

[3](a) Prof. Abdul Aziz, ICSSR Professor, National Law School of India University, formerly, Professor at the Institute for Social and Economic Change, Bengaluru; (b) Prof. M.K. Ramesh, Professor, National Law School of India University; (c) Prof. R. S. Deshpande, ICSSR Rajiv Gandhi National Fellow, Former Director, Institute for Social and Economic Change, Bengaluru.

The purpose was to ensure central focus, quality and improvements in the papers based on debates and discussions. The First Group Meeting was held in the Centre for Islamic Studies. Fifteen resource persons, well versed in Islamic scriptures—Quran and the Hafiz—participated actively providing valuable information to the paper writers. The Second Group Meeting was held at the Christian Association with eight resource persons participated in the meeting and providing inputs related to Bible, Old Testament and the New Testament. The current initiatives taken by the Church were also discussed in detail bringing out the relevance of religious institutions towards conservation and sustainable management of water. The Third Group Meeting was held at the Karnataka Sanskrit University with 25 resource persons participated in the meeting and providing inputs on the various perspectives of Hindu religion and its role in the protection of natural resources with a specific reference to water.

Work-In-Progress Workshops

We held four works-in-progress workshops on a monthly basis to review the progress of work. The review committee reviewed the papers in advance. During workshops, presentations were made by subject experts followed by reviewers comments which were based on the papers and the presentations made. Also organized a final workshop on the 'World Water Day', 22 March 2014 to which we invited a larger audience, resource persons across religions for their valuable feedback and suggestions. The papers were further revised based on the suggestions emerged during the final workshop.

At the very threshold of the enquiry, three scholarly efforts on Hindu, Islamic and Christian Philosophies, clearly emphasize and highlight, the primary and major distinction that has influenced the perception, prescriptions and practice concerning water. Water, as one of the five elements of nature (panchabhoota, namely, water, air, earth, fire and sky), is a manifestation of the divine being and hence sacrosanct, inviolable and invaluable under the Hindu faith. On the other hand, it is no more than one of the many creations of god—not the same as god—clearly apart, distinct as and of lesser significance than the Almighty, in the other two faiths. By that token, water is a resource, like any other, created by the creator. It can be objectified, commodified and evaluated in monetary terms and ascribed with an economic value. This idea and the fundamental difference has, indeed, steered the cultural orientations, ecological and economic understanding, use and management of water in the three faiths. Fascinatingly, all resource persons from all religions emphasized the significance of water augmentation, storage, protection, conservation, distribution and lessons to be drawn at all levels, for individuals and households, community and the state/government.

Based on an abundant literature on the above-mentioned aspects, there were several common aspects explained differently with a few aspects perceived differently as well. These ideas/perceptions have evolved over time and space with several factors having a profound influence, given the access to availability of water, which, in turn, influenced water resource management. For instance, in the Hindu scriptures, water is seen as being divine/holistic. However, Islam and Christian scriptures consider water as being created by a divine force and hence needs to be respected, protected and conserved.

Chapters

Given this backdrop, the chapters are organized as follows. This chapter provides an overview of how the messages from religious scriptures are relevant for sustainable water resources management and the process in which the scripture experts were engaged in writing the chapters. Chapters 2, 3 and 4 are contributions by scripture experts on water resource management in Islam, Christianity and Hinduism and Chap. 5 on legal perspective cuts across all the scripture-based chapters. Chapter 6 is on comparative overview across disciplines and the religious scriptures followed by Chap. 7 on policy perspectives that draw lessons for current contexts from across the scripture-based papers, respectively. To get an overall understanding, we have detailed them in the following paragraphs.

Chapter 2 'Water in Islam' by M. A. Siraj and M. A. K. Tayab refers the usage of references made on water in religious scriptures—the Holy Quran, Hadith and Sunnah and secondary sources—*Fiqh* and Islamic history. It is an earnest effort in presenting the Islamic scriptural directives and traditional practices related to water. The core aspects of the research work can be located in starting with 'Water in Holy Quran' and ends with 'Quranic Messages—Various Aspects'. These are faithful narrations of various Islamic Scriptural references, practices and cultural moorings over water, across dynasties. The references to ancient practices of irrigation, in particular, are informative and illuminating. Several aspects covered include judicious use of water, protecting water sources, the crucial need for storage of water, sustainable practices supported by guiding principles of water use, preservation and management, dogmas and practices regarding use of water in social and religious traditions which are well defined; technologies in water management are an interesting contribution as several innovative techniques evolved during the Mughal period and so on. Several verses are devoted to highlighting the ethical and moral messages of the Quran on the use and abuse of water which are very relevant and interesting.

Chapter 3 'Perspectives on the Water and the Bible' by Y. Moses refers to the Biblical records, which are inclusive of the Old Testament and the New Testament. The Biblical visualization of water as the source of life, as stated by the author, resonates with the Hindu scriptural conceptualization of water as life-giver and sustainer. The many splendored understanding of water is highlighted when the author draws attention to the over 700 references to water in the Bible. The passages cited and anecdotal references made by the author includes water is used as a metaphor for purification, blessing, relief and a force for rescuing from adversity and calamity.

There is a detailed account of the active engagement of the Christian institutions, in the education and propagation of the contemporary significance of the prescriptions and practices of the past that advocated access to and use of water, in a non-discriminatory manner, in meeting the basic needs. Water is seen as a spiritual symbol and metaphor with several connotations associated with the use of water in ritual cleansing and purification besides the recognition of rights of the poor, approaches to conflict resolution and so on.

Chapter 4 'Water in Hindu Scriptures—Thank you, Water' by Sudhakar Sharma and Shruti crafted a wide variety of scriptural citations and their interpretations in all their splendored glory. There is joi de vivre in the entire presentation—total involvement, reverential attitude towards the great traditions and a sense of gratitude for all the values one gets to learn from the past. Here, 'Water' becomes a metaphor for the divinity and the write-up is more of an ode and a tribute to water. Water, the concept and its myriad uses get described in great detail. The authors have drawn lessons from references made in Vedas[4] and scholarly manuscripts to several aspects of water management viz. technologies in water management and use, irrigation water management, water allocation and pricing, medicinal properties of water, and conservation of water for a sustainable use. Similarly, spiritual values associated with water have been pertinent in the ancient scriptures and practised in the daily routine rituals by communities. All this was perceived through viewing water as a 'divine component' where there was a greater emphasis on reverence for water, belief systems, religious practices through which they created a sense of awareness and respect towards water usage which helped conserve it as well.

Chapter 5 'Legal Lore of Water: Ecology and Scriptures' by Manjeri Subin Sunder Raj is a praiseworthy effort, for its sheer novelty and a welcome research exercise in hitherto an unchartered area, a wonderful canvas and a great starting point for future enquiries and analysis in law on Ecology, Scriptures and Water. In the legal conceptualization of water, a fairly adequate attempt has been made in bringing out the emotional, social, cultural and spiritual content and contexts of the times. The rights and duties, in relation to water and the treatment of the resource as a Common Property, are also discussed. The doctrinal bases of determination of rights and responsibilities over different manifestations and forms of water and water ecosystems are also given a due space for deliberation in this effort. Water governance and the principles anchoring them in different legal systems and civilizations have found a fairly good space in the enquiry. It also discusses the role of informal laws that played a major role in conservation of natural resources and the cultural practices across religions playing an important role, all these have a greater relevance to the current challenges particularly the violation of formal laws and linking the two in terms of drawing lessons for improved effectiveness and implementation in respect of the future water resources management.

[4]Vedas are the entire body of Hindu sacred writings, four books—the Rig-Veda, the Sama-Veda, the Atharva-Veda and the Yajur-Veda. Composed in Vedic Sanskrit, the texts constitute the oldest layer of Sanskrit literature and scriptures of Hinduism.

Chapter 6 'Water in Scriptures—A Comparative Overview' by K. V. Raju, S. Manasi, Abdul Aziz, M. K. Ramesh and M. S. Shruti discusses the evolution of water discussions across disciplines and drawing from the religion–scripture-based papers, an overview of common features that cuts across the rich cultures of these various religions. It highlights and emphasizes that all the scriptures across religions considered water with respect and reverence with slight variations in their perceptions about divinity, stewardship and others.

Chapter 7 'Policy Perspectives—Water and Scriptures' by A. Ravindra, K. V. Raju and S. Manasi discusses some of the current issues of water management, ecological sustainability and the emphasizes that the role of culture and learnings from scriptures would go a long way in addressing challenges in water conservation and management drawing some of the lessons drawn across the papers for policy that would be applicable to the current challenges in water resources management.

References

Alveré H (2012) Catholics for choice magazine, United States of America

Elaide M (1996) Patterns in comparative religion (trans: Sheed R). Nebraska Press, London

Gleick PH (2009) 'Thirsty word: Desperate quest for water', Excerpt from What Matters: The World's Preeminent Photojournalists and Thinkers Depict Essential Issues of Our Time, 20 Oct 2008. CNN.com

Moses Y (forthcoming) Chapter 2.2: water and scriptures—ancient roots and perspectives. Springer, New York

Nadkarni MV (2011) Ethics for our times. Oxford University Press, New Delhi, India

Nair KS (2003) 'Role of water in Development of Civilization in India—a review of ancient literature, traditional practices and beliefs' 'The Basis of Civilization—Water Science? In: Proceedings of the UNI-SCO/1 AI IS/I WI IA symposium, Rome, December 2003

Oestigaard T (2009) Water, culture and identity comparing past and present traditions in the nile basin region. BRIC Press, Bergen

Price M (2009) Changing minds to prevent climate change. Monitor on Psychology 40(9):16

Schelwald-van der Kley L, Reijerkert L (2009) Water: a way of life, Sustainable water management in a cultural context. Taylor and Francis Group, London, UK

Sharma KN (2000) Environmental Protection and Water Reverence in Ancient Indian Culture Secretary, International Commission on Irrigation and Drainage (ICID) 48, Nyaya Marg, Chanakyapuri, New Delhi, India. http://www.scribd.com/doc/16813661/Environmental-Protection-and-Water-Reverence-in-Vedic-Culture-by-KN-Sharma-Secretary-ICID

Siraj MA, Tayab MAK (forthcoming) Chapter 2.1: water and scriptures—ancient roots and perspectives. Springer, New York

UNESCO (2003) Water for people water for life. UNESCO report, World Water Assessment Programme

UNESCO Universal Declaration on Cultural Diversity (2002) A document for the World Summit on Sustainable Development, Johannesburg, 26 August–4 September 2002. UNESCO, Paris

United Nations (2006) Planet in Peril: an atlas of current threats to people and the environment. Minneapolis

World Water Day (WWD) (2006) United Nations Conference on Environment and Development (UNCED)

World Water Forum (2003) Kyoto, Shiga and Osaka, Japan

Websites

http://www.dictionary.com/browse/veda
https://en.wikipedia.org/wiki/Vedas
http://www.waterandculture.org/264_Water_Wisdom
http://environment.nationalgeographic.com/environment/freshwater/freshwater-crisis/
http://bora.uib.no/bitstream/handle/1956/3708/Book_WCI_lowres.pdf;jsessionid=D24DDC1A5E649A875611CE7670610873.bora-uib_worker?sequence=1s
http://www.apa.org/monitor/2009/10/climate-change.aspx

Water in Islam

M.A. Siraj and M.A.K. Tayab

Introduction

Life, food and water form one closely knit circular continuum. We could also visualize life to be poised/balanced on a tripod of '*air, water and food*', for it is only in water or with water that life processes can go on or sustained. Hence, clean potable water forms an indispensable component simultaneously; there is the output of waste products which are flushed out of organisms through water medium. These complex phenomena have given rise to studies such as ecology and environment. In this study, we shall trace the historical developments by way of contextualizing sacred texts and delineating the developments relating to the utilization of water. Thus, this chapter aims at understanding what these sacred texts as well as auxiliary texts have to say on these issues on the one hand, as also tracing their historical evolution, on the other. Hence, the crux is to relate how the people in different epochs solved their problems of getting adequate water and its efficient utilization and safe disposal of waste.

Methodology

This chapter bases itself principally on three sources, viz., the Holy Quran, *Hadith* and *Sunnah,* besides two secondary sources, viz. *Fiqh* and Islamic History.

Quran is a divine scripture revealed to Prophet Muhammad over a period of 23 years in the cities of Makkah and Madinah, where he lived during his prophethood. The verses of the Quran were brought by archangel Jibraeel regularly during (610 AD and 632 AD) and were memorized by the Prophet before imparting to a

M.A. Siraj (✉) • M.A.K. Tayab
Centre for Contemporary Studies, Bengaluru, India
e-mail: maqsiraj@gmail.com

K.V. Raju, S. Manasi (eds.), *Water and Scriptures*,
DOI 10.1007/978-3-319-50562-6_2

host of his companions who also memorized it through frequent recitations. The Prophet put the verses in a sequence (which is not the same as a chronological order of revelation). During the era of the third Caliph Osman, various versions that had been written down by that time were brought together, and uniformity was given to it. Ever since the days of Prophet Muhammad, people have been memorizing the entire 6000 and odd verses. Since the Quran is in rhymed prose, it lends itself to an easy memorization (called *Hifz*), which developed as a skill among Muslims.

Hadith (pronounced as *Hadees*) are the sayings of Prophet Muhammad, which come next in the hierarchy of religious text. There were great many narrators of *Hadith*, and they began to be compiled only after 200 years of the Prophet's death. By then, a lot of political changes had taken place with Islam graduating from a faith to an empire, while narrators had developed a variety of leanings and political affiliations. Consequently, *Hadith* with the same or similar contents could be found in varied forms and wordings by a host of narrators. Although six of the Hadith collections are known to be most authentic, i.e. *Sahih Bukhari, Sahih Muslim, Nisai, Tirmizi, Abu Dawud* and *Ibne Majah*, there is no guarantee of the accuracy and health of the text being original. According to the own admission of the six compilers, they did a lot of sifting, sorting, sieving and scrutiny, yet could not ensure that the wordings they were committing to writing were entirely those of the Prophet's. Besides, a lot of repetitions also bring down the actual volume of content that can be extracted from them. Thus, *Hadith* comes a distant second after the Holy Quran when it comes to accessing the Islamic teachings.

The third source is the *Sunnah* or traditions of the Prophet. This also forms part of *Hadith*. These are actions and deeds of the Prophet which his companions witnessed and related the account to others before they came to be recorded by the compilers of *Hadith*. These are purely in the wordings of the Prophet's companions.

Fiqh or jurisprudence is the fourth source for extracting the practices prevailing in the early Islamic societies. A bewildering variety of Fiqh has been produced by a large number of scholars. This is mainly their legal and juristic opinions on various issues related to the life and development in various societies, in various eras under a variety of political dispensations. Although it could be a guide to the way they elicited legal guidance from the actual scriptural sources of the religion, they may be totally irrelevant to our times. What is important is that, their opinion and formulae can be totally irrelevant today, but their method of arriving at a formula for their own societies and challenges typical to their times may be of some value for us.

Islamic History or Muslim history can be a source to know the way various Islamic societies and administrations adopted for dealing with issues of water management. Islam became an imperial force after 30 years of the Prophet's demise. The Prophet was succeeded by Abu Bakar (for tenure of merely 2 years), Umar (for 10 years), Osman (for 17 years) and Ali (for 4 years during which the Islamic state split between Madinah and Damascus). Umayyad Empire with its seat in Damascus and Abbasids with their capital at Baghdad became the principal centre of political administration. Later, the conquests brought lands up to Spain and Sind under their tutelage. Still later, a variety of sultanates came to rule over these lands with

Ottomans (from Istanbul), Mughals (from Delhi), Ayyubid and Fatimid (from Cairo), Al-Movahid and Al-Moravid (from Maghreb), carving up new sultanates. The history of these lands and dispensations has been documented by historians.

The references to the Quran have been provided in the brackets at the end of all verses (such as 2:14, i.e. 2 referring to the chapter and 14 referring to verse number). Since there exist a great variety of translations into English, in some cases, the name of the translator has also been mentioned.[1]

Water, a Natural and Precious Resource

Water is one of the most vital resources for man. Water is (along with food) an indispensable input for all living beings (animals or plants). Thus, water being such a vital resource, man should carefully and frugally use it, indeed conserve it. He should act as a conservator, caretaker and trustee of water.

Water, Its Uses and Disposal

There are numerous steps that have to be taken and they could be enumerated thus:

1. Getting or procuring water
2. Utilizing it as it is obtained
3. Water storage and its safekeeping
4. Augmentation of its usefulness (a) by frugal use; (b) cleaning so as to make it suitable for particular use; (c) a hierarchy of standards of cleanliness could be easily recognized, e.g. (i) water for injection should not only be distilled but also be pyrogen-free; (ii) water for drinking and cooking; (iii) water for washing and cleaning; (iv) for bathing and cleaning; (v) for gardening, animal husbandry, agriculture and aquaculture; (vi) for industrial processing and so on
5. Under item 4 above, another set of criteria needs consideration, (a) both optimal and minimal use; (b) conservation; (c) protection.

The above list is, of course, indicative of one of the tasks to be performed, desiderata to be kept in view and standards to be maintained. These are broadly the subheads under which we would like to proceed.

When we speak of the use of water, the question of a safe and benign disposal after use presents itself for tackling it concurrently.

Hence, gradation and initial separation of out flowing stream and its recycling become imperative for dealing with such a polluted water stream. A grading system helps classify and segregate various streams and decision taken to recycle them. The aim is not to let a raw, untreated stream into public water bodies. These may be lakes,

[1] The scheme of numbering verses of the Quran explained.

rivers or even the sea. A vigilant watch becomes imperative to prevent the polluter from escaping without paying towards the cost of reclaiming his waste stream. In a broader perspective, one can, of course, collect an element of cost towards this expense. In such a case, he does not have to bear this expenditure himself.

Rainfall, an Annually Renewable/Regenerative Process

Rainfall starts the cycle being the primary renewable source of freshwater. Rainfall varies according to the changing seasons during a given year. This annual cycle of rainfall helps replenish the supply of water to the parched earth. It is, of course, common knowledge that rainfall varies in intensity from place to place, season to season, etc. It is well known that a country like Malaysia or England receives rainfall for about 300 days in a year, whereas, at the other extreme, there are countries that receive rainfall for a limited number of days in a year, and what's more, even this may sometimes be copious or scanty. There are some areas hardly getting any rains during the year.

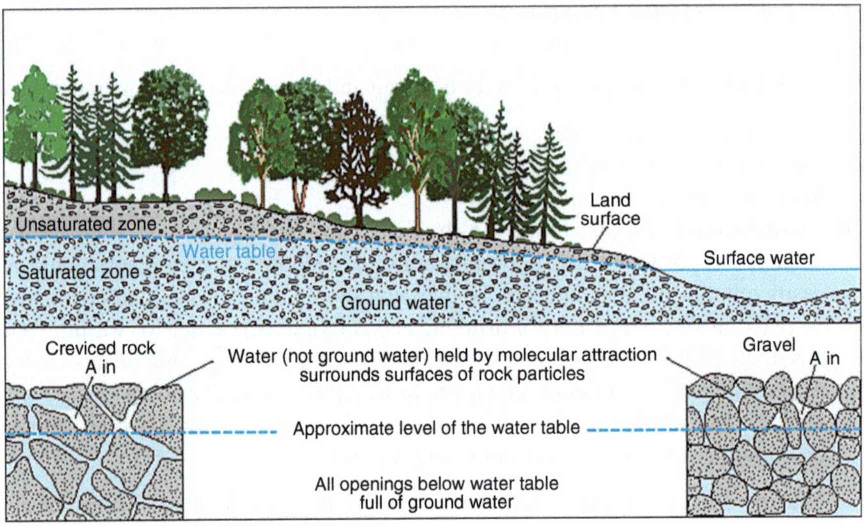

Surface and Groundwater Found in Nature

With the onset of rains, water starts flowing in small streams, rivulets, etc. They go dry when the rains are over. There are, of course, rivers big and small, seasonal or perennial. Apart from these sources, wells are also sunk which provide water either throughout the year or they may also go dry in summer. Thus, water is available seasonally, i.e. periodically, and a very few places are blessed with adequate water throughout the year. However, man and his livestock need water throughout the year for they cannot manage to survive without water for more than a day or two. Hence, there is clearly a disjunction between daily requirements and seasonal unavailability. This necessitates perpetual efforts by people to store water from any or all sources that could be tapped. They must store the liquid either by harvesting the rainwater or collecting from seasonal flows in rivulets, streams and groundwater. This struggle is common to all cultures and civilizations of: ancient, middle or modern periods. The quest for locating/producing food and ensuring an adequate supply of water becomes the very basic and irreducible minimum requirement for the very survival of mankind. We shall try to trace the trajectory of this quest for food and water. While doing so, we shall see how religious beliefs, cultural norms of behaviour and social mores have all acted and interacted in terms of shaping and moulding the direction and intensity of this quest.

Mapping Water on the Planet

Even before we proceed to map the human quest for water, it will be useful to have an overview of form in which water exists and its quantity and accessibility in the nature. It is estimated that 71% of the surface of the earth is covered by oceans and seas. It does not come as a surprise that 97.5% of the available water is found in the oceans. The salinity gradient, the action of winds, tidal activity, etc. power the engine of seasonal (climate) variations of the planet. Hence, the vast oceans no doubt form an essential component of the planetary system.

It is not as if there is no water on this planet. There is enough and more water. Nevertheless, it comes as a sombre reminder that 97.5% of this water is in the oceans. The remaining 2.5% is available in the form of freshwater. Out of this, as much as 85% is locked up in the polar ice caps, tundra regions and in what the geographers call the third pole, viz., the Tibetan plateau with its snow covered high mountains and glaciers. The glaciers store rainwater and release some of the rainfall that is received, as ice melt to supplement the water supply of rivers during off seasons. Glaciers play a vital role in moderating climate.

Thus, a tiny slice of less than 1% of freshwater is there for all the creatures of the land. As much as 70% of this freshwater is used for irrigating crop lands. However, it is estimated that by efficiently using irrigation water, consumption could be cut by one-third.

- Total water on earth comes to 1.4 billion km^3.
- Total renewable water falling on continents and islands/year is 41,000 km^3.
- World population is over seven billion today.

That is why somebody has observed that the reality is that there is essentially no more freshwater available on the planet today than there was 2000 years ago. It is indeed the same water that the dragons drank aeons ago.[2]

Freshwater being scarce is also the most distributed item on earth. 'Of all the planet's renewable resources, freshwater may be the most unforgiving, difficult to purify, expensive to transport and impossible to substitute, water is essential to food production…'[3]

Common Sources of Water

Having quantified the freshwater resources of the earth, we shall now look at the sources that are commonly tapped to get the required supplies. There are surface water sources which could be tapped easily such as a river or a lake. You may only have to dip your vessel and take the water. However, to tap groundwater, e.g. you may have to sink a well from a shallow open well to a deep bore well. The shallow wells tap the near subsurface water which is recharged during the annual rainfall cycle; thus, these are renewable sources.

However, the depth of deep bore wells may vary from 100 to 1000 ft. Of course, there are areas where they go down to even 10,000 ft! Here, water accumulation is not a result of annual rainfall. Water has accumulated over a geological period—meaning over centuries. It means tapping water from such sources amounts to mining water. When once exhausted, recharge will be a matter of hundred centuries or so! This has already happened in some arid zones. It shows how very careful and circumspect we should be while harvesting groundwater. It may be seen that we are refraining from using the common expression—'exploitation of groundwater'. To our ears, it is an absolutely inapt expression!

Quality/Grade of Water

We can distinguish between various qualities of water, e.g. water with some dissolved gas such as oxygen. Only such aerated water is suitable for aquatic life—irrespective of whether the creatures inhabit lakes, rivers or seas. That is why if water is boiled or steam is condensed, all the dissolved gases would have evaporated. No fish could live in such water. It is dubbed as sterile water.

---

[2] Earth is estimated to have, 1.4 billion km^3 of water. The three moons of the Jupiter have: Europa 2.9 times earth's (water), Callisto 27 times earth's, Gandymede 36 times earth's and among Saturn's moon , Titan has 29 times earth's water. Most of the above have it in the form of ice and some water underneath. [*National Geographic,* April 2010]

[3] C. P. Kumar: *Fresh water resources: A perspective*….Roorkee: International Year of fresh water 2003 www.angelfire.com/bc/mhhrrc/documents/fresh.html

Then, there is water with dissolved salt or sugar, for instance (or any other organic or inorganic substance). There is also saltish water, polluted water and so on.

Nevertheless, it may be appreciated that there is nothing like absolutely pure water for after all; water is well known as universal solvent.

Water, a Critical Resource

Water is a critically vital resource for sustaining life on earth; it behoves us to see what the Quran says about it. Moreover, even a cursory glance makes it clear that the Quran, the Bible and the Vedas, as do other scriptures and sacred books, all extol the role of water, while describing it extensively.

Quran Says All Life Created with and Dependent on Water

Quran is replete with references to the bounty bestowed on His Creation by sending rains…In the first instance, let's see two verses: 'God has created every animal from water....' (24:45). This verse is followed by another which adumbrates that 'It is He who has created man from water…' (25:54). This is followed by a third verse, which speaks about plants.

It is stated that 'It is He who sends down water from the skies; you drink thereof, and the plants also are satiated, on which you pasture your livestock, [and] by virtue of this, He causes crops to grow for you,…' (16:10 and 11).

Rainwater, a Bounty Showered by God

Further, about the quality and quantity of rains, 25:48 states that 'We cause pure water to descend from the skies…' and 15:22 adds that 'and we let loose the winds to fertilize [plants] and send down water from the skies and let you drink thereof…'

Sent as per Set Criteria

And in 23:18, it is stated that 'And we send down water from the skies in accordance with "set criterion" (measure) [set by us]…'.

The above extracts from the verses make it clear that water is essential for life on earth and is of course, necessary for survival/growth of plants, animals and man. It is iterated that rainwater is pure water and that it is sent down according to set criterion (measure). In view of these verses, the adage that without water 'we are nothing' rings true.

Thus, it is clear that man, from the beginning of history, has had to struggle and tasked to (a) find water; (b) having found it, he had to store it; (c) he had to keep it clean and (d) dispose of the waste hygienically.

A Myth That Persists

The economists of yore were fond of saying that water, air and sunshine are 'free goods' that nature provides to man. (Marshall). Even as late as 1890s, this was not perhaps an entirely incorrect depiction of the situation then prevailing. But that's not so now; freshwater (fw) is the most precious resource. It can no longer be considered a free good—indeed it's a scarce resource. Against this background, we may also mention that even at the beginning of the twentieth century, we had not known that other celestial bodies in the solar system possessed water. New evidence shows that mostly moons of some of the planets have water. (cf. footnote 1 below)

The Crucial Need for Storage

It has been mentioned above that (a) often water is not available throughout the year, and that all areas are not equally endowed with adequate freshwater (fw) though water is required throughout the year by both men and livestock. This calls for storage of water and if the habitation is in a rain shadow area or an arid zone, then water has to be brought from a distant source.

Men at the dawn of settled agriculture started building barriers (dams) across small rivulets and jungle streams. But they could build a small cross wall to store some water. This cross wall/barrier could be of modest height such that it was submersible when the stream ran high. It is known by several names in diverse places. However, in south India, it goes under the name 'Anaicut'. In ancient times, very rarely it could be of any great height. It was designed to be submersible. In the absence of any good impervious mortar, the wall would not be quite leak proof, and it was useful in storing limited quantity of water.

The next step was to use natural depressions in the topography for strengthening the sides of this excavation having a relatively big area as the foreshore of the tank/lake such that water could flow into this lake/tank-like structure. Later on, a surplus weir also came into vogue. For irrigating their lands, channels with a sluice/shutter were built.

It was in Mohenjo-Daro that the sharp-edged bricks began to be used in building water storage sumps. They were dry packed so precisely that there could be no water leakage. Over the millennia, similar bricks began to be moulded in most other civilizations.

Vessels made of burnt clay, hollowed out stone, (and in pottery) were widely produced that enabled people to store water at their homes. Skins also came handy not only to carry but also to store water. Temples in Egypt used vessels made of hollowed out marble, polished and decorated to store and offer sacred water to their deities.

Irrigation: A Legacy Handed Down by Ancient Civilizations

The use of irrigation as an aid to agriculture is an ancient art. It is indeed the very foundation, on which a civilization gets built. The probabilities seem to point to Egypt as its place of origin of organized irrigation. Evidence shows that the peasantry bailed up water at least 4000 years ago and there is evidence of earlier irrigation there and in other parts of the ancient world which strengthen the belief that irrigation preceded and made possible the establishment and development of ancient civilizations. Along with labour-saving devices such as Shaduf, Sakia and Archimedes screws were invented.[4]

Shaduf or paecottah: A long pole supported on a horizontal cross beam (8–10 ft from ground) was hung with a rope with the shorter end counterbalanced. When the other end had a bucket generally of leather was dipped in the river and when full, pulled with a rope. The beam was turned towards the receptacle of the channel where it was emptied. In Egypt, it was called Shaduf, and in India, paecottah. Depending on the depth of water in the river, these devices could irrigate about two acres of land. Before the mass availability of electric or oil pumps, it was a common sight to see men toiling away bailing water.

The Egyptian *Sakia* (called *harat* and Persian wheel in India) consisted of two parallel endless ropes passing over a wheel. Small buckets were attached and when the wheel was turned, one half of the small buckets got filled up, while the other half got empty. The main wheel was rotated by men or animals depending upon the head. This could irrigate more than double of previous device.

Archimedes screw: Then, there was 'Archimedes screw' invented circa 200 BC in Greece, in which a cork-screw rotated in a tube (wooden). When rotated, it bailed out water. 'Knowledge of these mechanical aids spread from one to other ancient empire. All these devices are still in use in various areas.[5] Trade and commerce greatly helped in such a dissemination.

Flooding or Basin Irrigation

Over a period of centuries, Egyptians studied—when the Nile starts having freshens, when floods come and when they subside and so on. With careful records having been maintained, they learnt to estimate the periodicity of these events.

[4] Encyclopedia Britannica, 1963 ed. P. 684.

[5] Ibid.

They would flood the fields when the river was in spate and let the water remain for the duration of the flood in the river. Then, the fields would be drained out and crop taken with the moisture retained being generally sufficient till the crop ripened. Up to nineteenth century, this method of flood or basin irrigation was in vogue in the Nile area. There are still one million acres under the old Pharaonic system of basin irrigation, while the remaining five million acres of Egyptian farmland being served by perennial irrigation, thanks to the dams built in the last century.

Whereas, the inflows in the Nile were more predictable and occurred more or less on schedule as compared to rivers in Mesopotamia where the Euphrates and the Tigris, originating in two different mountain ranges, behaved more unpredictably with floods being more severe. Even so, there also flood irrigation was in vogue by around 2000 BC. Hammurabi had ordered some canals to be cleared in 1700 BC.[6]

Mohenjo Daro civilization was based on and more attuned towards trade and commerce and was not a predominantly agricultural one. Hence, these irrigation methods were not adopted by them. Later on, in the Gangetic plain, all the above devices and systems gradually came to be used.

At the birth of Islam—in the Arabian Desert, water was, of course, one of the scarcest of resources. From ancient times, they had to be frugal and careful in using water. They depended on occasional oases with some scrub shrubbery around; sometimes, wells were also sunk in such spots. When Islam spread outside the Arabian Peninsula, they came in contact with areas such as Egypt, the fertile crescent and became familiar with irrigation practised in these countries. From the above, it could be seen that all these areas, Arab Peninsula, Egypt, Iraq and India where Islam spread, had had hoary traditions and practices well tried and tested, agricultural practices, and ways of irrigating their fields and husbanding their livestock and water.

Excerpts from the Quran

The Quran describes water as the basis and origin of life. Quranic verses point out (cf. p. 4 above) that life originated from water as the following verses amply show:

He sends down saving rain for them when they have lost all hope and spreads abroad His mercy (Quran 25:48).

Perceived Benefits of Water

Water's role in farming and producing the sustenance of life of men and animals is described in the following verse:

---

[6] Ibid.

He Who has, made for you the earth like a carpet spread out; has enabled you to go about therein by roads (and channels); and has sent down water from the sky. With it have we produced diverse pairs of plants each separate from the others. (20:63)

Yet another verse mentions the rejuvenating properties of rainwater hinting at the resurrection being a necessary stage in the cycle of life's ultimate destiny in the following verse:

And He who sends down (from time to time) water from the sky in due measure, and we raise to life therewith a land that is dead. Even so will you be raised (from the dead) (43:11)

A third verse hints at the purity of rainwater and we know for sure that the water pouring down from clouds is pure. This essentially points to the property of water which, even while dissolving the impurities into it carries them forward to seas and leaving them behind, rises into vapour to form clouds, before coming down to the earth again in pristine pure form.

Navigation over Water

Sea as Highway

In the previous paragraph, we have seen how water is crucial for the very survival of life on earth.

We now look at the use of large bodies of water as bases for navigation—whether it is seas or large rivers. Arabs, especially from the coastal areas were engaged in trade and commerce besides being intrepid seafarers. They built vessels for undertaking trade. In the Quran, there is a frequent mention of this facility. Here are some verses in this regard.

> 'And among His signs are the ships, smooth running through the ocean, (tall as hills)….' (42:32).

> 'It is God who has subjected the sea to you, that ships may sail upon it by His command, that you may seek His Bounty, that you may be grateful'. (45:12)

> 'Nor are the two bodies of flowing water alike—one palatable, sweet, and pleasant to drink, and the other salty and bitter. Yet from each (kind of water) do you eat flesh fresh and tender, and you extract ornaments to wear (e.g. pearls and coral), and you see ships therein that plough the waves, that you may seek (thus) of the Bounty of God that you may be grateful.' (35:12).

Prayer on Water

No Islamic prayer is prescribed in respect of water. But Muslims are recommended that they pray for rains whenever they face an acute scarcity of water and drought. This prayer is called Namaz e Istisqaa. This congregational prayer is held in an open ground with long supplication after the namaz (prayer) is over. The believers are supposed to beseech forgiveness for their sins and to invoke God's blessings for the inhabitants of the earth.

The Quran also narrates Prophet Moses' prayer for water which led to gushing forth of 12 springs from stones upon hitting them with his staff:

And (recall) when Moses prayed for water for his people, so we said, 'strike with your staff the stone'. And there gushed forth from it 12 springs, and every people knew its watering place. 'Eat and drink from the provision of Allah, and do not commit any mischief on the earth, spreading corruption'. (2:60).

Qualities of Water

Fertilizing quality of water: The Quran talks of water's quality of rendering the soil fecund and thereby paving the way for agriculture:

> 'And we let loose the winds to fertilize [plants] and send down water from the skies and let you drink thereof…' (15:22).

Mythological Anecdotes

Water is held as a precious resource to be made accessible to every individual. However, the water which carries high sanctity among Muslims is the water of Zamzam. It springs from a well in Makkah, 20 m east of the Kaaba, the holiest place in Islam. According to an allegorical description, it was a miraculously generated source of water from God, which began thousands of years ago when Ibrahim's infant son Ismail was thirsty and kept crying for water and was kicking at the ground when water gushed out. Millions of pilgrims visit the well each year while performing the Hajj or Umrah pilgrimages, where they reverentially drink its water. Most of them carry a few litres of Zamzam water while returning from their pilgrimage.

It is also seen that some of the pilgrims dip their *Ihram* (sacramental two-sheet cloth wrapped around the body) in Zamzam water. This is brought back after drying and kept safely to be used as a burial shroud upon death, while laying them to rest in the grave.

The fervent faith of pilgrims, from early Islamic times to this day, has not ceased to lend curative properties to the water of the Well *of Zamzam,* which is perceived as a bearer of *baraka* (blessings) as well as cure. It is stressed that Holy Prophet, peace be upon him, drank this water and used it regularly for various purposes.

Promoting Conservation and Protection

Sustainable Practices: Guiding Principles of Water Use, Preservation and Management

A host of laws, regulations, principles, practices and norms have either been laid down or deduced from the scriptural sources or evolved over centuries in Islamic lands regarding use, storage, preservation, distribution and management of water and water resources.

There are three basic principles that inspire and inform these regulations and practices:

1. *Tawheed* (unity of God and his creations)
2. *Mizan* (It is the principle that every aspect of creation holds together because there is a balance. Maintaining natural systems, therefore, can be interpreted as *mizan.*)
3. *Khalifa*: Human beings have been created in the role of vicegerent of God on the earth and have been entrusted to use, manage, preserve, promote, conserve and distribute the resources. These resources are *amanah* (trust) with the man appointed a custodian.

Islam forbids its followers from wastefulness, overconsumption and hoarding of resources. Water is regarded as a common property. Traditions of the Prophet Muhammad describe a common share for all people in respect of three things: (1) grass (pasturelands), (2) water and (3) fire or fuel.

In view of the crucial nature of water in sustaining life, God has made its use the common right of all living beings. All are entitled to its use without establishing any monopoly over its resources. Similarly, its usurpation, despoilment, wastage and abuse have been strictly forbidden. Regardless of the use being private or public or in circumstances when it is scarce or the supplies are abundant, believers have been forbidden from extravagance. Once the Prophet saw his companion Saad washing before his prayers, he admonished him against wastage of water, telling him that even if he were to be washing on the banks of a river, he should not waste water.

As for sharing of water, the Prophet is said to have decreed that no more than ankle depth of water be taken for irrigation. In later ages, in Muslim lands, penalties came to be levied on misuse, pollution and degradation of water.

During his time, the Holy Prophet demarcated two types of zones (1) *Harim* for the sake of protection and (2) *Hima* for preservation.

Hima were zones where water wells, springs or streams or watersheds were located. These were to be reserved for the collective use. Everyone would have access to them but no one would use them for bathing or washing or any other kind of activity that would pollute or harm it in any way.

Harims were inviolable sanctuaries and were larger areas reserved for growing of trees and woodlands. The animals would be allowed to browse and graze. No trees would be cut nor wildlife disturbed.

Hima and *Harims*, present till very recently in many of the Islamic lands, are no longer used as water has come to be a civic responsibility of the city administration.

A Hima in Sanaa, Yemen Still in Use

Purity and cleanliness of water received a great deal of attention in the Prophet's teachings. He forbade people from passing urine in or near a water source as also taking bath in stagnant water. The Prophet is quoted to have said: Guard yourself from three kinds of practices: No one should evacuate his bowels near water sources, on roadsides and in the shade.

Awareness Through Belief Systems

Conserve, Do Not Splurge

Islam would like its adherents to be circumspect in the use of earth's resources. There prevailed a strong awareness and common concern among people with respect to taking care of water sources and surroundings at the time of Islam's advent. However, the development of concern regarding ecological degradation is a development of the second half of the twentieth century. But the very rise of Islam in desert lands amid scarcity of water and fodder had in itself engendered an attitude of extreme caution towards the use of water. Islamic teaching as a general rule, urges its followers to be prudent consumers of resources. It enjoins austere lives, charity instead of conspicuous consumption. The Quran says: Eat and drink out of what God has provided, but do not indulge in waste. For those who indulge in wasting resources are indeed brothers of Satan. The Quran says:

> And render to the kindred their due rights, as (also) to those in want, and to the wayfarer: But squander not (your wealth) in the manner of a spendthrift. (17:26)

In keeping with these specific instructions, household utensils were patterned for economic use of water. For instance, the ewers (see picture), jars and *badhna* from Muslim lands came to have nozzles in order to pinpoint delivery of water. This is distinctly different from *Lota* in households in India. Similarly, Leather skin bags (*Mushk*) that could preserve water from evaporation were in common use in Islamic lands. Moreover, since most of the lands where Islam won adherents in the early period were basically desert lands, water had to be carried on the back of horses, mules or donkeys. The leather skins filled with water seemed a more practical mode of transportation for they could be laid on horsebacks with men striding on them.

Water Purification Practices (Past and Present)

1. Rainwater is of the highest purity. So people were always given to collecting it as it was free from contamination. But the storage capacities being limited, the availability of collected rainwater for drinking and cooking was limited as it could last for a couple of days at the most.
2. Well water was normally quite potable provided the particular well was not surrounded by organic waste and other debris.
3. Water from rivulets, streams, rivers and tanks was the next source.
4. Keeping water from the above sources (2 and 3 above) in a vessel and letting it stand for some time helped the particulate matter settle down. This process was hastened by adding alum. Some locally available plants and herbs such as Velliver or Khas also came in handy which were put in the vessel to clean water. All these steps helped remove gross particulate matter, as also coagulate some organic matter. But when water was exposed to organic waste including human and animal waste, it became problematic. The density of population, adequate availability and scarcity of water were also crucial determinants of the extent of such contamination in a particular habitation. In fact, dense population in a particular habitation in itself proved to be a major cause for a high degree of pollution of water.

However, sand filtering by way of putting sand into a vessel and keeping another vessel below to collect the slowly descending water was a common practice handed down from antiquity. A more elaborate method was to use four vessels one over the other with the bottoms of three upper vessels having some small holes in them. The first vessel contained small pebbles, the next one charcoal and the third vessel half to two-third of coarse sand. After sometime, a thin organic film would form helped entrap organic matter in the water. Thus, in the bottom vessel, clean filtered water got collected.

During the last century, boiling of water came into vogue. However, it may be noted that these methods were mostly practised by the well-off persons, while the poor could hardly practice it on a continuing basis.

At present, various methods are being practised especially by the municipal authorities for cleaning/purifying water. All these are cultural practices adopted in diverse climes and regions.

Recent developments have shown that a bottle full of water kept exposed to sun for a day or two effectively kills off all germs and viruses in water and makes it safe for drinking purpose (as also for cooking). Thus, getting potable and safe drinking water is a constant struggle.

Use of Water in Social and Religious Traditions

Dogmas and Practices

Islam does not allow/encourage worship of any tangible object. Since Islam is a strictly monotheistic faith and images of God are prohibited—be they drawn, painted or sculpted—Islamic mode of worship does not involve any tangible object or material. This rules out the scope for any kind of prayer or supplication being fashioned around water. However, ritual cleansing with water is a mandatory component of Islamic prayers regardless of their being said at home, at mosque, etc. This ritual cleansing is termed *wudu* or ablution. Next comes a bath.

Need for ablution led to mosques reserving an area for ablution with provision of cisterns, series of taps and overhead tanks. Mosques constructed in olden days in Iran and the subcontinent had *hauz* or open cisterns with seating arrangements around them. Usually, the cisterns would be raised a feet above the ground in order that the user could scoop out water from cupped hands and wash the face, hands and legs over a spillway that would carry the outwash away from the ablution area. Some older mosques had Hammam or bathrooms attached for travellers. These were mostly copied from Turkish inns and mosques and came to the subcontinent via Iran.

Prophet Mohammed established a religious institution (declared as collective property) known as 'waqf'. In this context, some of the water sources and wells were declared as 'waqf' and public had the right to use them. Generally, Islamic principles concerning water laws were based on two norms. These were (1) the right of thirst where all people had the right to quench their thirst or to give water to animals and (2) the right of irrigation where water can be used to water land and plants.

In the Shei'i school of Islam, the concept of serving of water to wayfarers or processionists is considered an act of piety. This led to the establishing of 'watering stations' or Saqqa Khana for free serving of water for drinking, and it came to be recognized as a religious duty. The Shi'ites endeavour to honour and commemorate the tragic event of Karbala and the martyrdom of Imam Husain and that of his loyal companions, who were martyred after remaining thirsty for 3 days. These Saqqa Khanas could be seen adjacent to Imambaras in Lucknow, Murshidabad and Jaunpur, cities with sizeable Shia population in India. One such modern Saqqa Khana was constructed at the famous Imambara at Jaunpur in 1990 where water is dispensed electronically.

Resources Enough for Man's Need

In verse 41:10, it is stated that.

'[He has].... Blessed it [i.e. the world] with plenty and growth, and ingrained the means of growing its food within it, sufficient for all seekers...' (Translation by Ahmed Ali). We discuss this issue further in paragraphs below.

Respect to Rivers in Scriptures and Mythological Descriptions

As stated above, certain norms were laid down to ensure that rivers were kept clean. In certain places, any infringement attracted penal provisions under the law.

Water Festivals of the World

Islam does not prescribe any festivals related to water, rivers or rainy season. However, in Mughal period, *Phoolwalon ki sair* got evolved as a syncretic practice. Phoolwalon ki sair has its origin to the reign of penultimate Mughal King in Delhi, Akbar Shah II (1808–1837). It was started by his wife Mumtaz Mahal Begum who took a vow to offer a chador of flowers at the dargah (mausoleum) of Khaja Bakhtiyar Kaaki at Mehrauli if her son Mirza Jehangir was released by the British from internment. When this happened, the entire Mughal darbar shifted to Mehrauli for a week and a floral chador was offered at the dargah and a floral *pankha* (fan) was offered at the temple of Yogmayaji. Seeing the enthusiasm of the people, the Mughal King decided to make it an annual affair immediately after the rainy season. It continued up to Quit India Movement in 1942. The British discontinued it under their 'Divide and Rule policy'. Independent India's first Prime Minister, Jawaharlal Nehru, revived it in 1961 and attended it at Mehrauli till he lived. Now, it has grown into a huge affair with pankhas being taken out in procession and dance troupes putting up performances. Chief Minister of Delhi attends it regularly and occasionally. Prime Ministers too have been putting in an appearance. All states of India offer floral pankhas for the purpose to both the dargah and the temple. It is organized by Anjuman Sair e Gulfaroshan (Society of Festival of Florists), a registered society. The message of secularism is embedded into it, and the fact that it is held after rainy season is considered a festival of thanksgiving for God's bounty.[7]

Technologies in Water Management

Awareness Regarding Water Cycle and Hydrology

Water being a very crucial element in sustaining life, the Quran amplifies the idea of the primacy of water. It also delineates how water is replenished in the nature. Some of the verses of the Quran shed light on how winds, clouds and rain complete the water cycle. Here, we quote two of them:

> Allah is the One Who sends forth the winds which rise up clouds. He spreads them in the sky as He wills and breaks them into fragments. Then thou seest raindrops issuing from within them. He makes them reach such of His servants as He wills. And they are rejoicing. (Quran 30–48)

> Allah is the One Who sends forth the winds like heralds of His Mercy. When they have carried the heavy-laden clouds, we drive them to a dead land. Then we cause water to descend and thereby bring forth fruits of every kind. Thus we will bring forth the dead. Maybe you will remember. (Quran 7–57)

[7] www.phoolwalonkisair.com

Against the then popular notion that groundwater owed its existence to the ingress of ocean water towards the interior of the continents,[8] the Quran provided the following interpretation:

> 'We sent down water from the sky in a (definite) measured manner and lodged it in the ground. And we certainly are able to withdraw it. Therewith for you we gave rise to gardens of palm-trees and vineyards where for you are abundant fruits and of them you eat'. [Quran 23–18 to 19]

Traditional Sustainable Practices

Abstemious Use of Water

The general instruction regarding water was to minimize its use and conserve it. The Prophet is reported to have used not more than one *mudd* (approximately two-thirds of a litre) of water and two to three and half litres of water for *Ghusl* (bath). There were instructions not to waste water even if one were to take bath on the bank of a river.

Architecture and Designs for Water Storage and Conservation

Development and widespread use of Aqueducts (Qanat) or underground canals was a major contribution towards conservation and transport of water over long distances by Muslim rulers through centuries. Although these were not originally devised by Muslims, they learnt it upon arrival in the Nile Valley and were responsible for its widespread replication in the valleys of the Euphrates and the Tigris and still later in the Iberian Peninsula (Spain).

[8] G. Gastany and B. Blavoux, *Encyclopedia Universalis,* Thales of Miletus's theory as quoted by Hydrogeologists.

The Qanat system is one of the most ecologically balanced water recovery methods available for arid regions because a Qanat relies entirely on passive tapping of water table by gravity, without upsetting the natural water balance.

The Qanat system consists of underground channels that convey water from aquifers in highlands to the surface at lower levels. Qanats exploit the natural gradient of the land to transport water underground to the agricultural and urban areas below.

The conduits—which are usually 50 to 80 cm wide and 90 cm to 1.5 m high—vary between several 100 m to more than 100 km in length. In Iran alone, there are some 22,000 of them, comprising more than 273,500 km of underground channels.

The spread of Islam initiated another major diffusion of Qanat technology. The early Arab expansion helped establish Qanats across North Africa into Spain and Cyprus.

With major Islamic empires having had their sway over terrains which generally experienced harsh climate, several Muslim rulers experimented with these underground canals (aqueducts) to preserve water as well as use the gravitational flow for carrying it over long distances.[9]

Nehar e Zubaidah

Zubaidah (died 831 AD), wife of Baghdad ruler Haroon Al-Rasheed (ruled 170 AH–193 AH), realized the great scarcity of water in the holy cities of Makkah and Madinah when she herself went on a pilgrimage. She ordered the engineers and surveyors to locate some water sources in the vicinity of these places. A large spring of water was found in the Valley of Hunain on the outskirts of Madinah. She ordered that a canal be dug (roughly 500 km) on the pilgrim route to take water to the pilgrim sites of Arafat, Mina and Muzadalifah. She funded the entire project. Surveyors recommended that open canals would not be viable as harsh climate would result in evaporation of water. An underground aqueduct was recommended and water stations for pilgrims were constructed at frequent intervals for travellers to collect supplies. Parts of the Zubaidah canal can be seen even today by those who perform Hajj pilgrimage. The system was functional till some 35 years ago and supplied water to the areas. It suffered heavy damage in the form of floods in Numaan Valley and became defunct in 1400 AH (37 years ago).

[9] Mohammed Reza Balali and Jozef Keulartz, *Technology and Religion*, The Qanat underground… irrigation system, (www.academia.edu).

Burhanpur Model

The Mughals and other Muslim rulers built water supply systems in several Central Indian towns that carried water through underground aqueducts to cities and habitations. Two such systems still survive in Burhanpur and Aurangabad where 'qanats' (literally underground canals) originated from a well-like vertical shaft in mountains. These used the Persian hydraulic technology. 'Kundi Bhandara' as it is known in Burhanpur (Indian state of Madhya Pradesh) is part of such a system which even to this day supplies clean potable water to a population of about 50,000 in the city.

The system was laid by Mughal governor of the area Abdul Rahim Khan in 1615. Seven different tunnels like qanats—broad enough to allow a man walk standing inside—emanating from the shaft in Satpura Hill ranges carried water to the city. On the way, water also seeped down from the earthen walls, thus adding to the original stream. Burhanpur is said to be a garrison town hosting an army of 250,000 with the entire supplies coming through the qanats. There were two advantages, i.e. preventing the loss from evaporation as well as averting any possibility of poisoning from the rivals. It is said at its peak, the system generated 100 million litres of potable water a day which is on par with what the City of Burhanpur needs today for a population of 300,000.[10]

Water Management Under Adil Shahi Dynasty

Adil Shahis of Bijapur were expert managers of water. They conceived water as a luxury commodity and not merely as an essential life resource. The water was collected from surrounding hills and brought to the city through tunnels and supplied to bavadis (or bowdies as the wells are known in the South Indian plateau) in the city.

[10]Sudhinder Sharma, *Deccan Herald*, 'A Timeless Hydrological Wonder Allowed to Go to Seed, January 4, 2013).

Ali Adil Shah built 16 lakes and talabs in and around the city during his reign, with some of them surviving even to this day. Among them, Begum Talab is considered a technological wonder of its age. Built in 1651, it is situated two miles to the south of the city. It could contain 25–30 million cusecs of water. It had been planted with trees all along its embankments to reduce evaporation and had lotus plants floating all over. Lotus is said to have purifying properties. In one corner of the talab, there is a room made of stone (now full of mud) which is said to be the room from where water flowed to the city through underground conduits. The conduits were made of earthen pipes leading the water to gunjs (underground cisterns in the localities). Gunjs are water towers ranging from 25 to 40 ft where water stood at a constant level. Many of the Gunjs are still working, collecting water.

Alongside these talabs, they also created bavadis or tanks. A few prominent among them are Taj Bavadi, Chand Bavadi and Ibrahim Bavadi. Others include Nagar Bavadi, Ibrahimpur Bavadi, Mantra Bavadi, Alikhan Bavadi, Langar Bavadi, Basri Bavadi, Hashampeer Bavadi, Jansa Bavadi, Sandal Bavadi, Bukhari Bavadi, Mubarak Khan Bavadi, etc. One Captain Sykes, who visited Bijapur in 1851, documented 200 bavadis and 300 kuwan (wells) inside the Bijapur Fort. These kuwan or wells were all interlinked in order to ensure that even the last well had an ample supply of water. Nearly 30 bavadis are serving the people's need for water even to this day. Some of these bavadis are tourist spot today for the grandeur of their architecture.

A similar network of aqueducts used to tap water from Hursul and Kham rivers flowing down from the mountains surrounding the town of Aurangabad in Maharashtra. The system was pioneered by Aurangabad ruler Malik Amber in 1617 AD as the town faced an extreme scarcity of water during summers. These were called 'Nehar' or Canal. More aqueducts came to be constructed in the later centuries, with the last of them coming in 1803 AD. It is said that the one constructed by Malik Amber was large enough to allow a man astride a horse conveniently negotiate all through its length up to the city. Two canals among the latter additions were Shah Mehmood Canal of Panchakki and Shah Ali Nahri. Shah Mahmood's canal brought water to the Baba Shah Musafir's Dargah in the heart of the town. It was constructed by Turktaz Khan in 1695 AD, a noble in the court of Nizamul Mulk Asaf Jah. It channelled the water from a spring through to the Dargah and was used to run a flour mill for grinding the grain for a garrison. Another stream of water was cascaded over the pool in front of the Dargah complex which comprised a serai (inn), a mosque and a madrasa. The water was used for cascading till very recently.

In a desert-like terrain, carrying water in such kutcha (unmetalled) canals exposed to the skies was not worth the effort, while for seepage and evaporation loss could be very high with a little of scarce water reaching the users' end. Hence, they had to have covered canals to carry water. The side and bed of the canals had to be made as impervious as possible, given the technology and available materials. An aqueduct is usually an elevated structure/canal for water to pass from one valley to the other. Otherwise, those canals were laid in the field in covered form. Arched elements were used for construction.

Canals Dug During Sultanate and Mughal Rule

During a 650-year Muslim rule over Delhi, the Sultans (belonging to six different dynasties between 1191 and 1528) and Mughal emperors (from Babur through to Bahadur Shah Zafar between 1528 and 1857) mainly concentrated on digging of wells and constructing of canals across rivers emerging from the Himalayas. Since land revenue was collected in kind, increased agricultural production would mean a better collection. In order to boost production, farmers were provided with loans during the reign of Muhammad bin Tughlaq (1325–1351). His father Ghiyasuddin Tughlaq dug a canal for promoting agriculture.

Ghiyasuddin Tughlaq (1320–1325) was the father of Sultan Muhammad Tughlaq. The credit for creating the biggest network of canals before the nineteenth century goes to Sultan Firuz Tughlaq (1351–1386). He constructed canals across various rivers in the north, viz., Yamuna, Sutlej, Ghaggar, Kali, etc. These canals brought in lot of irrigation water to the farmers.

The Mughal era witnessed several renovations and a further expansion of canals. Emperor Akbar renovated the Yamuna canal in 1568 originally constructed by Firuz Shah Tughlaq. However, the Mughals, the great lovers of garden that they were, added another objective to digging of canals, i.e. to supply water to gardens and fountains in their palaces. Yamuna Canal, repaired during Akbar's reign was extended further to Red Fort (Lal Qila) during the reign of Shah Jehan. He also ordered the digging of a canal from River Ravi through the Shalimar Garden at Lahore.[11]

---

[11] P. G. Shinde, paper presented at the National Level Conference in 2005 on Water Management Scenario 2025 Problems, Issues and Challenges.

Aesthetic, Decorative and Scientific Use of Water

Water did not stand just for its use in agriculture, irrigation and industry-related and domestic needs, but also had other connotations. Under the Quranic metaphors, water symbolizes paradise, righteousness, and God's mercy. This influenced the use of water as a motif in the Islamic architecture quite frequently. From numerous references to cooling rivers, fresh rain and fountains of flavoured drinking water in paradise, one can deduce that water is the essence of the gardens of paradise. It flows through them, bringing coolness and greenery, besides quenching thirst. The believers are rewarded for their piety by 'rivers of non-stagnant water and rivers of milk, unchanging in taste and rivers of wine, delicious to the drinkers, and rivers of honey purifying' (Quran 47:15). The water in paradise is never stagnant; it flows, rushes, unlike the festering waters of hell. The Quran also equates the water of paradise with moral uprightness; 'in the garden is no idle talk; there is gushing fountain'. (88:11 to 12).

Pools were used to act as mirrors enlarging the visual space of the monuments. Channels and pools were the most important elements of Islamic gardens. Originally necessary for irrigation, they were later developed for their visual beauty besides being incorporated into elaborate architectural features. Within the palaces of the Mughals and their Rajput allies, and other ruling dynasties, variations in the systems of copper pipes carrying water for cooling terrace pavilions, channels flowing through royal chambers, fountains and water-gardens, and underwater collection tanks were the norm. Thus, here too, various water-storage methods were devised, as were a range of water-lifting mechanisms. While in religious buildings, water was mainly used for ritual cleaning, in palaces and houses, it was meant to provide coolness and décor.

Persian wheels, using the force of water for lifting water to great heights for irrigation, came to be developed in various centres of Islamic civilization such as Cordova, Grenada, Seville (all in Spain), Iran and Syria. These devices were installed across naturally flowing streams or waterfalls.

Introduced in India by the Mughals was the most advanced water-lifting device. On the Persian Wheel, a garland of pots was attached to the rim of a wheel, with a gear mechanism also attached to it. Sadly, not all the water-related architecture of Mughal India has been fully documented. Therefore, there is an urgency to do so before this unique feature of the land's hydraulic past is lost in the face of a rapid shift of water management to the municipal authorities and the destruction of many old buildings and sites!

Bowlies at dargahs: Most dargahs, mosques and Imambaras had *bowlies* (*bowdies* in South India) dedicated to the general public. Over time, the devotees associated healing properties with such water sources.

Water-Lifting Machines

When it comes to hydraulic engineering, the name of Badi al-Zaman Al-Jazri (1136–1206) should not escape a mention. Popularly referred to as Al-Jazri in the scientific literature, Badi al-Zaman was a polymath combining in him skills of an inventor, mechanical engineer, craftsman, artist and mathematician. He lived in Jazirat ibn Umar, currently known as Cizre (pronounced as Jizre) in Eastern Turkey. He is best known for his book *Al-Jami bain al-Ilm wa al-amal al-nafi fi sinaat al-hiyal* (The Book of Knowledge of Ingenious Mechanical Devices, in short *Al-Hiyal*) which he completed in 1206 in which he described 100 mechanical devices. His forte lay in producing mechanical devices rather than just theorizing them.

A-Jazri invented five machines for lifting water as well as watermills and water wheels with cams on their axle used to operate automata. Some of the devices were:

Saqiya chain pumps: Al-Jazri's saqiya machines used the crankshaft in a chain pump for the first time. He also built a water-raising saqiya chain pump which was run by hydropower rather than manual labour, even as the Chinese were using hydropower for chain pumps before him. Saqiya machines, like the ones he described, supplied water in Damascus since the thirteenth century up until modern times, and were in everyday use throughout the mediaeval Islamic world.

Double-Action Suction Pump

Al-Jazri invented a twin-cylinder reciprocal piston suction pump. This pump is driven by a water wheel, which drive, a system of gears, an oscillating slot-rod to which rods of two pistons are attached. The pistons work horizontally opposite cylinders, each provided with vale-operated suction and delivery pipes. The delivery pipes are joined above the centre of the machine to form a single outlet into the irrigation system. This water-raising machine had a direct significance for the development of modern engineering. This pump was remarkable as it was the first known use of a true suction pipe (which sucks fluids into a partial vacuum) in apump and could lift water up to 13.6 m.

Water Supply System

Al-Jazri developed the earliest water supply system to be driven by gears and hydropower, in thirteenth century Damascus for supplying water to its mosques and Bimaristan hospitals. The system had water from a lake turn a scoop-wheel and a system of gears which transported jars of water up to a water channel through which water flowed to mosques and hospitals in the city.

Varied Uses of Hydropower

Al-Jazri used hydropower for varied automated machines such as automatic gates driven by hydropower. He also created automatic doors as part of one of his elaborate water clocks. He also invented water wheels with cams on their axle used to operate automata. He also devised an automated flush mechanism for hand-washing, peacock fountain with automated servants, drink-serving waitresses, a musical robot band, also called 'a musical automation', which featured four musicians on a boat that floated on a lake to entertain guests. Prof. Noel Sharkey observes that it is quite likely that it was an early programmable automaton.

Water-Powered Elephant Clock

Al-Jazri devised a water-powered elephant clock which was installed in the courtyard of a mosque in Damascus. The timing mechanism was based on a water-filled bucket embedded inside the elephant. The bucket had a bowl floating in water, but with a small hole in the centre. It took half an hour for the bowl to fill up through the hole. In the process of sinking, the bowl pulled a string attached to a see-saw mechanism in the tower on top of the elephant. This released a ball that dropped into the mouth of a serpent, making the serpent to tip forward thereby pulling the sunken bowl out of water via strings. Simultaneously, a system of strings made a figure in the tower raise either the left or right hand and the mahout to hit a drum. This indicated a half or 1 h. With the gong being hit, the snake would slide back to its normal position. The cycle then repeated, as long as balls remained in the upper reservoir to power the emptying of the bowl.

A working replica was built for Ibn Battuta Mall in Dubai, United Arab Emirates, while another could be found in the Musée d'Horlogerie du Locle, Château des Monts, in Le Locle, Switzerland.[12]

Water clock: A water clock was constructed based on the Archimedes Principle by Ridwan bin Al-Saati in 1203 and was positioned at the gates of the Umayyad Mosque in Damascus.

Water Use and Management in Islamic Civilization

Rights to water got evolved over a period of centuries in the Islamic lands. Generally, three rights came to be recognized universally. For example, (1) everyone had a right to drink or slake his thirst; (2) all pets, cattle and household had a right over water; (3) right to irrigation. This established the hierarchy of use, i.e. first humans, followed by animals and farmers for irrigation.

The first two rights, i.e. quenching one's thirst and giving water to animals were generally referred to as *Chafa* in the Islamic law or laws in the Muslim lands. The third right, i.e. using water for irrigation was known as *Chirb*. Based on these two universal rights, the nature of water (i.e. groundwater in wells, springs and surface water such as found in lakes, ponds or community cisterns, or running water like in rivers) also led to laws being formulated for regulation of water supply in detail. However, following a major split between Sunnis and Shias, there were significant differences among them regarding the ownership, pricing and sale of water. Yet it must be borne in mind that even today 90% of the Muslim population falls within

[12] www.muslimheritage.com, http://en.wikipedia.org/wiki/Elephant_clock

the category of Sunnis, while the remaining 10% follows Shiite Islam. Among the 56 member-states of Organization of Islamic Cooperation (OIC)—an apex organization of Muslim countries currently, excepting Iran, Iraq, Azerbaijan, and Bahrain, all other states are dominated by Sunnis. The general principles that guided water management in the Muslim lands are outlined here:

1. Everyone had a right to access wells or water points dug for quenching one's thirst.
2. In case of scarcity, animals could be water fed after everyone has had one's share of water.
3. In the case of nomads, they had an exclusive right over the wells dug by them for the entire period of their stay; however, they were not to refuse anyone the right to slake his/her thirst.
4. After the departure of the nomadic community, the wells were to become public property with users having access to it on first-come, first-served basis.
5. Anyone who sunk a well on his own land or unoccupied land was the sole owner of water and was not required to supply water irrigating others' lands.

The Ottoman Empire which governed a vast part of the Muslim world over three continents of Europe, Asia and Africa till the beginning of the last century had codified the Islamic law pertaining to water extraction, storage, protection against pollution, maintenance and operation of waterworks and distribution of water. However, the element of conservation and augmentation was not much in evidence in this law codified around 1870s, but left a strong legacy in a majority of the nation-states that sprang out of the empire after 1921 when Turkey jettisoned all its colonies and became a secular nation-state. Broadly, the law as laid down in codified form in Mejelle had the following broad guidelines:

1. All waters were declared vested with the state, or crown, or incorporated in the public domain, with the state thus taking over the place of the Muslim community.
2. Every use of water (other than for drinking or watering the animals) left free under the colonial legislation or sharia was placed under the government control.
3. Water commissions were set up to survey and recognize established water rights.
4. Land registers were compiled for keeping a written record of duly recognized land and water rights.

For the legal status of water, the following points from the *Mejelle* (Book of codified law) could be pointed out[13]:

Article 1234 of the *Mejelle* defines water as an unsalable commodity to which everyone has a right (*mubah*, which means permissible). Groundwater as well belongs to the community (Article 1235).

---

[13] Dante A. Caponera, Principles of Water Law and Administration : National and International, Balkema, Rotterdam, 1992.

The definition of water as non-saleable, public owned commodity applies to running water which has not been appropriated to water contained in the wells dug by unknown persons and to waters of the sea and large lakes (Article 1236).

Water for drinking and for feeding animals watering (*hakki chefe*): Everyone may quench his thirst from both privately and publicly owned rivers.

Article 1268 provides that 'anyone who has on his property a stream, basin or well may prevent any other person from trespassing into his land to obtain water except for drinking in which case, the user should not damage this property, i.e. the rim of the well or the sides of water conduits.

Everyone is entitled to use water for irrigation from lakes or streams forming part of the public domain, provided the rights of the third party are not infringed upon. Irrigation canals and ditches as well as pumping installations can be freely constructed. However, nothing should be done to cause floods, exhaust water supply or lower the water level in as much as the passage of a boat is hindered.

The use of privately owned waterways is restricted to the riparian landowners. A non-riparian landowner will have to obtain the consent of all riparian landowners in the event of his using it.

The sale of rights of way, of irrigation rights and overflow of water from conduits, is permitted as part of the sale of land. However, if the owner of the land sells his land together with the right to draw water from a certain river or a canal, third parties having rights to that water of the river or the canal, benefit from the right of pre-emption (Article 1015). The landowner may, however, sell his land without the water rights.

Rivers falling in the public domain will be maintained by the State. However, expenses can be recovered from the users (Article 1321).

Private waterways will be maintained by the owners who are entitled to draw water for irrigation or for their animals (Article 1322).

All co-owners were responsible for maintenance of water courses in proportion to the benefit they derive from it (Article 1326).

The Harim (inviolable zone) around the well is fixed at 40 *arshuns* (a cubit, corresponding to 0.758 m) where the trespassing is prohibited (Article 1281).

Harims around all kinds of water bodies (rivers, springs, underground, etc.) were laid down under Articles 1282–1292.

A well dug over one's own property (*mulk*) will have no harims. And the owner of the adjacent property too will be entitled to dig a well on his property and the former will have no right to object.

Differential Rates of Wealth Tax

Zakat or wealth tax figures among the five basic duties a Muslim must fulfil during his life. Zakat was levied at the rate of 2.5% of the accumulated wealth (i.e. if its value exceeds 88 gm of gold and if it has remained with an individual for over a year). Zakat on agricultural produce was termed *Ush'r*, or tithe. It is derived from the Quran which says:

> O Ye! Who believe! Spend of the good things which ye have earned and of that which we bring forth from the earth for you… (2:267).

The later jurists opined on the basis of the practices of the Prophet that *Ush'r* would be collected at the rate of 10% of the total harvest if the land was watered by the natural streams, springs or rivers. In case the land was irrigated by artificial means such as wells and buckets, Ush'r would be collected at the rate of 5% only.

Perhaps, the wisdom behind it was to maintain a difference in the tax structure between those farmers who benefitted from natural irrigation and those who invested efforts in arranging water for crops. While not too much should be read into it, it can be taken as a hint for rebate for those spending on organization of their own watering arrangements. Alternatively, it could also be interpreted as asking the beneficiaries of natural resources to cough up more by way of taxes.

Over time, Muslim communities and nations developed the norms and laws regarding ownership and distribution of water from natural sources. For instance, rivers (the Nile, the Tigris, the Euphrates and the Jordan River) were to be treated as belonging to the entire communities (read nations in the later avatar). Everyone would have the right to benefit from them, and no one would harm them. Still later, these were supplanted by the international convention of riparian rights, now universally recognized.

Muslim theologians as well as jurists divided water sources, for the sake of ownership, into wells, water springs, rainwater streams, large and small rivers, human dug canals, and irrigation channels. Generally, three categories were known:

1. Water in personal containers, treatment plants, reservoirs or distribution systems to be considered private property. Owner will have the right to own, use, trade and sell it.
2. Water sources situated on the privately owned lands such as ponds, tanks, streams and springs will have a restricted right to others. Other's rights will accompany their obligations towards them.
3. Rivers, lakes, glaciers, aquifers, seas, snow-fed streams and rainfall will be considered public property or what is in today's parlance 'Commons' and cannot be traded, sold or withheld from someone. However, those who invest in raising infrastructure or extracting it will have certain rights. (This area of legislation was not very comprehensive to fulfil today's needs. All Islamic nations today recognize the Convention on the Law of the non-Navigational Use of International Water courses which elaborately lays down upstream and downstream rights, regulations regarding the preservation of ecosystems associated with them, pollution and penalization, etc.

Quranic Message—Various Aspects

Enjoy the Sustenance Provided

As we have observed in the previous paragraphs, the advice having been given to avoid behaving outrageously, we are now being asked to enjoy the sustenance being provided. This may come as a surprise to some persons that the Quran does not advocate asceticism, but indeed it urges people to partake of the bounties provided by Him and enjoy the sustenance thus provided and of course be grateful to God. Wide ranging moral and ethical injunctions are laid down.

2:172, O you who have attained to faith! Partake of the good things which we have provided for you as sustenance, and render thanks unto God, if it is [truly] Him that you worship.

77:43 Eat and drink to your heart's content: for that you would have worked (righteously). While consuming food and drink bestowed by God, do not forget the poor—your fellowmen.

6:141… Eat the fruit when it comes to fruition, and give [unto the poor] their due on harvest day. And don't waste [God's bounties]: verily He does not love the wasteful!

Maintain Balance

This is an omnibus heading—sort of a hold-all reference point—for the word 'balance' is quite a frequently used term in the Quran. 42:17 reads, ' it is God [Himself] who has bestowed revelation from on high, setting forth the truth, and [thus given man] a balance [wherewith to weigh right and wrong]'.

A similar idea is expressed in 57:25… 'Revelation… bestowed a balance [wherewith weigh right and wrong] so that man might behave with equity',

In Surah 55, verses (7, 8 and 9) state that … '[He] has devised [for all things] a measure, so that you [too, O men!] might never transgress the measure [of what is right]: weigh, therefore, [your deeds] with equity, and cut not the measure short!' A reading of the above section makes it abundantly clear how this matter is emphasized in [57:25], 'Certainly We sent our apostles with manifest proofs, and we sent down with them Book and the Balance so that mankind may maintain justice…' (Translation by A. Q. Qarai)

42:17, it is God who has sent down the Book with truth and balance.

28:77, By means of what God has given you, seek the abode of the hereafter, while not forgetting your share of this world. Be good [to others] just as God has been good to you, and do not try to cause corruption (*fasad*) in the land.

Be Frugal in Use, Waste Not and Pollute Not

Some of the Quranic verses lay emphasis to state '… Eat of the good things that we have provided for you, and be grateful to God…'

Simultaneously, the Quran asks us not to waste (water, food, etc.). It states … 'But waste not by excess for God loves not the wasters' 6:141.

The excerpts from these verses show (there are several more) that God enjoins us to eat as also enjoy whatever bounty of food and drink has been bestowed on us. But they clearly ask us not to be wasteful and extravagant.

The responsibility not to waste is squarely cast on human beings—thus each one of us has to honour/obey it.

Don't Behave Heedlessly, Oppressively or Tyrannically

You should not consume heedlessly, i.e. far in excess—in a word, do not be a glutton. Be moderate in consuming food [water and other resources]. Don't *splurge*. Don't do it to show off so that people may see how well off you are!

Further, it is stated (5:87), 'O you who believe! Make not unlawful the good things which God has made lawful for you, but commit no excess: for God does not love those given to excesses'.

Prophet Shuaib warns his flock against behaving tyrannically. (26:130 to 134). As these verses lay down: 'And whenever you use force and become tyrannical; So be conscious of God and follow me; [who] Gave you increase of cattle and sons, and orchards and springs. I fear the punishment of a terrible day for you'.

Referring to Pharaoh and his tyranny the verse says (28:5), 'We wished to favour those who were weak in the land and make them leaders and heirs'.

Those who oppress and terrorize in the land unjustly are warned of painful punishment.

The concept of oppressed, i.e. of those whose rights are trampled upon by the rich/powerful forms the theme of several verses. We are citing only two of these verses, here water—a scarce resource is more often than not commandeered by the rulers, the elites, the well-off, i.e. rich and powerful while the poor/weak are denied or deprived of access to water or at least left to fend for themselves. The two verses being cited are 4:98, 'But those who are helpless, men, women and children who can neither contrive nor formulate a plan nor do they know the way [out of their predicament]'.

Such utterly helpless people would complain (4:75) 'What has come upon you that you do not fight in the cause of God and for the oppressed men, women and children who pray : Get us out of city, O Lord, whose people are oppressors, so send us a friend by your will, and send us a helper?' (Translation by Ahmed Ali) (The oppressed are addressed in the Qurn as *mustadhafun*).

42:42 'blame attaches but to those who oppress other people and behave *outrageously* on earth, offending against all right: for them is grievous suffering in store'

(Asad). Asad explains that (in footnote 44) 'although this is primarily a reference to those who oppress [other] people and behave outrageously on earth, offending against 'all right', the meaning of the term is general, applying to all kinds of deliberate evildoers.

The Ethical Moral Message of the Quran on Use and Abuse of Water

Some Sociological Aspects of the Distribution of Water

Verse 41:10 states that '[He has].... Blessed it [i.e. the world] with plenty and growth, and ingrained the means of growing its food within it, sufficient for all seekers', (Translation by Ahmed Ali).

This verse explains that God has blessed all those who seek to satisfy their thirst and hunger as also to satisfy related wants, with adequate means. If we look around carefully, it could be seen plainly that there exists a symbiotic relationship between man and his surroundings or environment. Before settled agriculture, i.e. in the pastoral and nomadic times, the population of earth was hardly in millions. As agriculture developed and its intensity increased, it could progressively support more and more people so much so that now the population is estimated at around 6–7 billion souls! However, the ghost of Malthus never seems to really disappear. It appears to be always looming in a corner, warning humanity that it has exceeded its welcome on the planet.

Nevertheless, humans inhabit almost all the climes: from cold Arctic zones to hottest deserts, to lands abundantly endowed with water to water-scarce areas. Men living in relatively comfortable zones may be wondering why some men are living in such inhospitable zones?

Apart from this, the innate capacity present among men varies to an astonishing degree those who have acquired more knowledge, those who are physically more dexterous, or those who are plainly more aggressive come to acquire/possess more power. A majority of them do exercise it to their advantage at the cost of the poor and the weak.

As a result of the above factors, everyone does not enjoy equal access to and use of water. Its corollary is the inability to earn/get food. Thus, we see one paradox upon another paradox. In India, e.g. physically, the minimum quantity required is available, but not all the poor have an equal access to water as also food. We have so far been viewing the side of availability of water. In this and following sections, we would like to look at the distribution side of water. How are consumers faring? How easy or difficult is for them to access their water requirements. Let's take two or three examples by way of elucidation. This, in a nutshell, becomes an issue for justice and equity in the distribution of scarce resources, particularly water.

It is quite frequently found explained in the Quran that it is a 'Book of Guidance for mankind' (*Hudan-lin-nas*). This guidance is permeated by an overwhelming emphasis on ethical–moral behaviour in whatever activity one is pursuing.

The first section below sets out the Quranic injunctions and their message. The subsequent sections apply these injunctions in respect of water.

15:85 'we Created not the heavens, the earth, and all between them; but for just ends…' (Translation by Yusuf Ali).

Having created the earth, God has given it to man in trust to make best and judicious use of it. Simultaneously, He has granted/endowed him with power to think and the ability to choose between right and wrong. God assures that He will reward those who are righteous. But those who choose to go astray shall be punished for their wickedness. Naturally, those two abilities distinguish him from all other creatures of the world.

In 16:90, it is laid down: 'Behold, God enjoins justice, and the doing of good, and generosity towards [one's] fellow men, and He forbids all that is shameful and all that runs counter to reason, as well as envy, [and] He exhorts you [repeatedly] so that you might bear [all this] in mind'.

Next, it speaks of different endowments granted to men in 35:32 that 'we have bestowed this divine writ, as a heritage unto such of our servants as we choose: and among them are some who sin against themselves, and some who keep halfway [between right and wrong], and some who by God's leave, are foremost in deeds of goodness: [and] this indeed, is a merit most high!'

In a tone of astonishment, it is queried, 'Do they not reflect in their own minds?' 30:8.

Here is the call to behave equitably and justly. Verse 4:135 enjoins that 'O you who have attained to faith! Be ever steadfast in upholding equity, bearing witness to the truth for the sake of God, even though it is against your own selves or your parents and kinsfolk. Whether the person concerned be rich or poor, God's claim takes precedence over [the claims of] either of them. Do not follow your own desires, lest you swerve from justice: for if you distort [the truth], behold God is indeed aware of all that you do!'

The message as regards the importance of behaving morally is reiterated with three elements being mentioned in 11:85, 'So, O my people, weigh and measure with justness, and do not withhold things due to men and do not spread corruption in the land, despoiling it'.

Justice and Equity

One of the major and recurrent themes in the Quran is that of justice and equity. This emanates from one of the attributes of God that is justice. This question has been discussed in great detail by Daud Rahbar in his seminal work, 'God of Justice'.[14]

We have stated in the foregoing section that man has been endowed with the power of choosing between good and evil. The problem takes off from here. Man's

[14] Daud Rahbar, *The God of Justice*, Brill, Leiden, 1965.

self-interest or even selfishness manifests in unfair and unjust behaviour towards his fellowmen.

We have so far seen the various aspects of water in terms of its availability or scarcity and the various sources from which it is obtained and so on. This entire endeavour is naturally to see that all living beings get their water requirements at least in terms of a basic minimum level.

*Justice (***Adl***)*

Here, we recount three verses of the Quran which speak of justice and equity.

5:8 'O you who believe, stand up as witnesses for God in all fairness, and do not let the hatred of a people deviate you from justice. Be just: This is closest to piety: and beware of God. Surely God is aware of all you do'. (Translation by Ahmed Ali, later AA)

7:29 'My Lord has enjoined justice …' (AA).

In another instance, the Quran enjoins in 49:9 '…Deal equitably [with them] for verily God loves those who act equitably'.

Trusteeship

Man a Trustee, Not a Despoiler

The scheme of creation as outlined in the Quran may be summarized thus. In 2: 30, it is stated, 'and Lo! Thy Sustainer said unto the angels: 'Behold, I am about to establish on earth one who shall inherit it'. Asad explains that 'establish on earth a successor or vicegerent (i.e. the *Khalifa*). Derived from the verb *Khalaif*, which means succeeded, {another} it is used in this allegory to denote man's rightful supremacy on earth. It is most suitably rendered by the expression 'he shall inherit the earth (in the sense of being given possession of)'. In some verses, human beings are described as *Khalaif al Ard* (2:30 and footnote 20). In 35:39, it is stated 'He is who has made you inherit the earth'.

The earth has been endowed with vast resources, which can be used—utilized by man to satisfy his myriad wants. In this vein, there are several verses in the Quran.

Here, we can recall the adage, life, food and water, form a closely knit continuum (page 1 above). Food and water are the basic minimal requirements for survival of all forms of life. Besides this, as the awareness of man increases, so do his wants. To satisfy them, he has to explore, test, devise and manufacture commodities. It may easily be seen that food and water occupy the apex position so to say of his wants.

The hierarchy from the point of view of the needs of human beings could start from food and water followed by raw materials for production of various commodities and so on.

The basic approach taken is that these resources are placed at the disposal of man and he is to act as a 'trustee' of the resources; as such, it is incumbent on him to use them circumspectly, carefully and frugally. He should not wantonly spoil or damage these resources.

Several of these issues are discussed in the following sections. It may be appreciated that all these resources in the form of water, food and all other raw materials are finite in quantity as part of the earth's system. Except for the input of solar radiation, which comes from outside the earth, everything else originates in earth and thus finite. We have seen above that, of water available on the planet, only 1% or so is freshwater. It is the same water that the dinosaurs drank several millennia ago. When we view the problem from the overall planetary system, it becomes quite clear that there is no scope for man to behave irresponsibly, wantonly and mindlessly, especially when utilizing water (as also all other resources). This is the reason why the Quran repeatedly calls man as a trustee of such resources. We may emphasize that some of these resources are vast, but nevertheless finite. They are most emphatically not infinite.

Equity and Trusteeship

We have seen above that the Quran lays emphasis on the concept of trusteeship of wealth. What's more, this is being stressed without any ambiguity. God is the Creator of the earth, sun and moon, indeed the entire universe. It is also laid down that the wealth that is legitimately acquired by dint of hard work is declared as *Halal* earning and that the wealth acquired through deception, deceit, aggression, etc. does not fall under the *Halal* (permissible/approved) category. Although in general parlance, a person of such wealth is called its owner, it may be noted that the possession itself turns out to be quite transitory or ephemeral. He is in temporary possession of it and that as a matter of fact, he cannot arrogate to himself the ownership thereof. Thus, he can enjoy the usufructs associated with it for the time being. In the Quran, he is called the trustee of such wealth. That he should meet his obligations towards charity, be compassionate towards the poor and the have-nots. These principles are enshrined in the text, however, the praxis is far from this ideal, in that these ideals are heard and understood, but not followed in practice. As a result, so far as water is concerned, the rich/powerful continue to enjoy a much higher share than what they are entitled to all through these centuries.

If you look at the situation prevailing in the twentieth century, we find that it was a time of considerable tumult when a number of ideologies were in sharp conflict with each other. Capitalism, socialism and communism were competing with each other. At this juncture, there was the voice of Gandhiji who advocated four essential principles of his philosophy. They were truth, ahimsa, trusteeship and constructive cooperation.

Gandhi said trusteeship was a means through which the wealthy people would be trustees to the wealth and look after the welfare of people in general. Such a view was condemned by socialists as being in favour of landlords, the feudal princes and the capitalists. Gandhi believed that people could be persuaded to give their (surplus wealth) to help the poor. This is how the Mahatma put it, 'Supposing I have come by a fair amount of wealth—either by way of legacy, or by means of trade and industry—I must know that all the wealth does not belong to me, what belongs to me is the right to an honourable livelihood no better than that enjoyed by millions of others. The rest of my wealth belongs to the community and must be used for the welfare of the community'.[15]

As Stiglitz puts it, capitalism moved towards a situation akin to robber baron era of the nineteenth century USA. The financial meltdown was the result of a quite suspicious, nay dishonest activity of the financial sector.

Ecological Balance and Pollution

We have seen above that the Quran states that whatever has been created by God is according to a plan, executed as per the blueprint. Everything that goes into making a thing strictly follows the prescribed measure and is well proportioned. It is thus fine-tuned to perfection. In other words, it is truly balanced in terms of several dimensions. When such a balance and equilibrium is disturbed or deformed, we could call it that the substance/thing has been degraded or polluted.

This issue is dealt with in 30:41 which reads as follows ' Pollution has appeared in land and sea because of what man's hands have wrought, so that He may let them taste something of what they have earned, in order that they will turn back [in repentance]'. Here, pollution refers to the disturbance/disruption wrought by the activities of human beings. That has led to—from mild to serious consequences. They have to face the consequences of their misdeeds. If they take a remedial action immediately, the intensity of consequences may be either mitigated or eliminated.

In this verse, the word used is 'Fasada'—a polysemic word. Hence, several meanings have been derived from this word and the verse. It is often translated as corruption, imbalance and disturbance from equilibrium, etc.…here *fasada* has been translated as pollution. It can be viewed as disturbance from an ecological balance point of view. Let us now see what explanation is offered by M. Asad for this verse in his Chapter 30, footnote 39, 'thus, the growing corruption and destruction of our natural environment, so awesomely—if as yet only partially—demonstrated in our times, is here predicted [predicated too] as an outcome of what men's hand have wrought, i.e. of that self destructive—because utterly materialistic—inventiveness and frenzied activity which now threatens mankind with previously unimaginable ecological disasters: an unbridled pollution of land, air and water through industrial and urban waste, a progressive poisoning of plant and marine life, all manner of genetic malformations in men's own bodies through an ever-widening use of drugs and seemingly "beneficial" chemicals, and the gradual extinction of

[15] Trusteeship of M. K. Gandhi, compiled by Ravindra Kumar, 4/68.

many animal species essential to human well-being. To all this may be added the rapid deterioration and decomposition of man's social life, the all-round increase in sexual perversion, crime and violence, with, perhaps, nuclear annihilation as the ultimate stage: all of which is, in the last resort, an outcome of man's oblivion of God and, hence, of all absolute moral values, and their supersession by the belief that material "progress" is the only thing that matters.'

Problems Associated with Commons

We have cited Marshall who in his (Principles of Economics) holds that sunshine, air and water are free goods provided by nature to man (page 3 above). We have stated there perhaps in the 1890s and early 1910s, this might have been not an entirely untrue view to hold. But in today's situation, freshwater (as also fresh air) has become a scarce resource indeed.

Those goods and sources used and enjoyed by everyone generally can be referred to as 'commons'. You may have, e.g. a road, brook, beach and mountain scene. Everyone can enjoy it both individually and in the company of others. However, if road is found unpaved/kutcha path, on which not much expenditure has been incurred initially, nor is there a heavy maintenance expenditure involved, its use may be quite free. However, if the road has been built involving a heavy investment by the State or any other entity, and costs quite a bit for maintaining it, it may be necessary for the user of such a pucca road to contribute towards its cost by way of toll.

If water were to flow in a jungle stream before accumulating into a natural depression, i.e. a small tank, then the public can certainly use it without paying any toll. But if a reservoir is built across a river with a considerable initial investment and it costs to maintain it, the question that arises here is: an agriculturist taking water for irrigating his field, should he get it (a) entirely free of any charge; (b) -meet the cost partially; and (c) meet the full cost? Similarly, if water from this reservoir is piped to a city where it is filtered, stored and distributed, could a system of piped delivery to homes be free or the consumer should pay (a) some nominal cost; (b) a modest bill; (c) meet the total cost? We have no scope to enter into elaboration of these options. But suffice it to say that one can certainly go to a river, and drink water to his fill without paying anything, but to get the same water through a tap in his house, a considerable cost has to be borne to access it. Here, conceptually, water is free, but delivery at the 'end of the pipe' has to be paid for by the consumer. This principle has to be understood by all those who clamour that water is a free good that nature provides—why should we pay for water at the tap? In view of the above explanation—such a view is clearly naïve or uninformed one.

We may thus state the premise that water is of course a free good provided by nature. But if it has to be stored, cleaned, filtered and piped to your house, you may have to pay a price for it. It becomes a priced commodity, far from being a free good. So a free good so modified emerges as a priced good. This gives rise to the question of pricing, cross subsidy, etc.

In this case, there are several stakeholders, but we will take up the most relevant of them: the consumer (individual), the community and the State.

Naturally, the individual may like to pay the lowest price. The community is expected to take a broader view in terms of willingly paying more than what the individual would like to pay. Here, the State should play the role of a moderator or arbiter. The issue should be discussed among the three stakeholders thoroughly and they should strive to come to an amicable outcome. If they find themselves unable to come to any workable solution, it will be a recipe for chaos which should be avoided at any cost.

If such a tangle were to arise, the aforesaid principle of justice and equity, of fairplay—guidance could/should be drawn from textual sources, and they should strive to solve the problem faced by all the stakeholders. The principle that natural resource like water is a bountiful gift from God and that they are only trustees should not be lost sight of. The trustees should act morally to keep his private interests in check as a fair-minded operator (this applies to all the stakeholders mentioned above).

Here, rules and regulations properly worked out through a consensus and promulgated by the State should go a long way towards an equitable solution to problems that might arise.

Water Is No Longer a Free Good

Water is a basic need of life. The question whether it should be available to all as a free good or marketed as a commodity has cropped up in the wake of neoliberal economic liberalization. Urbanization has brought in new demands that have changed the mode of thinking vis-à-vis the needs of life. Freshwater is no longer accessible as it used to for people inhabiting villages normally around tanks, river banks or community wells. Although governments, under the obligation of making the basic needs of life accessible to all, have invested hugely in building up a basic infrastructure of water supply network, electricity, housing and roads, are finding it hard to maintain them due to low and unsustainable returns. Hence, much of the cost of infrastructure gets subsidized only to be extracted from citizens by way of taxes on other services that the State renders.

Supplying water to millions of people inhabiting our megalopolizes entails pumping water from distant sources. Electricity to pump this amount of water itself costs 30 to 50% of what most cities spend on their water supply. A huge network of pipes, overhead tanks, underground reservoirs, valves and siphons have to be maintained for building up reserves and maintaining supply schedules. Often a basic cleaning of water at the source and protection against pollution require proper monitoring.

Not merely supply, even the collection of used and wastewater through an intricate network of gutters, sewers, and storm water drains has also emerged as the concomitant civic responsibility to ensure hygienic conditions and public health.

So, many of mediaeval ethos, norms and regulations regarding water use may not be simply applicable to the modern times. Even the basic premise, i.e. whether water should remain a free good, is begging a reexamination in a setting where 'free' is becoming synonymous with 'poor quality, unreliability and, to some extent, 'risky'. If huge investments on water were to remain bereft of prospects of any returns, or returns not adequate enough to sustain, what hope is there of guaranteeing their quality and reliability in the long run. Similarly, if it were to be priced, what yardsticks are to be followed by a society so very stratified? Could there be differential rates for the use of the precious liquid in urban homes which range from using water merely for basic cooking, washing and drinking to villas that employ it for car-washing, lawn-watering, swimming, musical fountains and for fish ponds. Could water supply rates be fixed proportionate to the floor-space area, inmates of premises, income levels of households, etc. need to be addressed.

Learning for the Future

Development in Near Future

Large dams: Twentieth century saw the era of mega dams. The slogan was bigger the better, e.g. Tennessee Valley Authority and Colorado dams in the USA, similarly, large dams came to be built in the then Soviet Union including those which sought to change the course of the rivers flowing towards Arctic circle to drain into the middle of the Soviet Union and their fiasco. In India, Damodar Valley Authority, Bhakra Nangal, etc. were built. Multipurpose dams—generating hydro power and supplying water for irrigation flood control was the rage during those days after independence. They held that bigger was better and larger dams were synonymous with growth and development. That resulted in not so benign consequences and sometimes even serious ones. This has resulted in a full scale reappraisal of the theology of large dams.

Nevertheless, the shortsighted greed of the State and of the people in taking up or planning a very large number of small dams in a known quake prone area like the Himalayas is very unnerving a prospect. We had a foretaste of the catastrophe that was likely to happen when these projects were under way in Uttarakhand and Arunachal Pradesh.

In this connection, we come across a number of warnings and admonitions cited in the verses.

RO System

A human body can be thought of as a series of pumps—big and small; a series of valves and above all membrane filtration systems. Taking a leaf from this, a remarkable development took shape during the last half of the century. From being a lab curiosity, reverse osmosis (RO) process has emerged as a successful industrial process for turning salty to highly brackish water into potable water. It is a process which is scalable from 5 L a day to 5 million litres a day output! As compared to water→steam→water process—the energy input is extremely favourable in RO. Further improvements to reduce the cost are, of course, constantly being made. Arid areas in the USA and the Middle East have huge facilities for RO. Households in India are big users of RO. This is indeed a most welcome development in providing good potable water where none was available before.

Root-Zone Irrigation

The other development in this vein is root-zone irrigation or drip irrigation which is especially crucial to water hungry crops. This has proven to be extremely useful in areas with a high degree of scarcity of water.

Industry Needs to Economize on Water Usage

In industry, water is mostly used as a mass–transfer medium by various industries including chemical processing industries. Water was free once upon a time or cost very little. A lot of work is under way to economize use of water, but more important breakthroughs could be expected only when there is a change of mindset among the captains of industry.

A breakthrough towards in situ desiccation of human excreta in combination with solar power could be done economically. It would dramatically reduce the consumption of freshwater as a mass–transfer medium besides reducing the pollution load too…several interesting developments is on the anvil.

This is a very short recounting of the possibilities and developments that are going on in various fields.

May we be permitted to close this section with the fervent hope as follows?

16:8 ‘and [it is He who creates] horses, mules and asses for you to ride, as well as for [their] beauty: and He will yet create things of which [today] you have no knowledge’.

Asad inter alia—explains this by saying let’s God will create new things of which you have no knowledge [today] is valid for every period—past, present and future’.

Salient Points That Emerge

1. Earth’s useable water resources are finite and represent only 0.5% of water on the planet.
2. We are making excessive demands on this stock of water by way of using, polluting and depleting it at a rate much more than the nature’s capacity to replenish it.

3. Stockholm Water Symposium held in 2000 declared that 31 countries are facing water stress and scarcity with over a billion people having no access to safe drinking water. It also urged the world to recognize the terrible reality that by 2025, two-thirds of the world population will be living with acute water scarcity.
4. Humanity in the past was mainly guided towards a frugal use of water for its daily needs more due to inaccessibility to water than its unavailability. Technology was the main constraint in accessing water sources.
5. Although nature's conservation was not so much of a concern then, the religious saints and scriptures with their dominant sway over people motivated them to be modest in the use of water, instilling a fear of accountability to God as also inculcating a sense of sharing of earth's resources with fellow human beings.
6. With complicated waterworks awaiting development and advent of relevant technology, religious leaders, in their own humble way, created common facilities such as public wells, cisterns, saqqa Khanas or bowlies at Dargahs for access to and availability of water to the common folk. Perhaps, this could be the only way to ensure an equitable access to and availability of water to people. It implies the recognition of water being 'a common good' with no authority and control over it. The *kuwan* (well) and *bavadis* (step well) etch to relief the different roles these water bodies were to play in public life. While wells were generally considered meant for private use, the step wells by their distinct architecture implied that everyone was welcome to benefit from them including animals free of cost. It was essentially a usufruct model which allowed everyone to derive profit from these water bodies without claiming any right over them, or causing any damage to or depletion of its supplies.
7. If not exactly sacred, water in Islam was considered 'a common property' with the religious prescriptions recognizing the right of everyone when it came to slaking one's thirst and water feeding animals even from private wells. Islam recognized the role of human beings only as a trustee of Natural resources, not as owners or proprietors.
8. Finally, it was only in very well-developed civilizations such as Ottoman Empire or Mughal rule that Muslims could conceive of building large water storages and a public distribution system as discussed under Burhanpur and Aurangabad models.
9. A cursory look at the hydraulic technologies developed/promoted by Muslims suggests that they largely catered to the premises occupied by the elite and were not capable of democratizing the water supply as is done today.

The Threat of Globalization

Meanwhile, globalization has come to dominate the world economy during the last quarter century. It aims at rendering water, just as any other natural resource like mines, forest and land, a private commodity to be sold and traded on the open market. They no longer view water as a social resource essential to all life forms on the planet earth, but an economic resource to be managed by market forces, akin to any other commodity. While in public discourses they are not tired of talking of privatization of water services as being socially beneficial, in practice, it is the economic imperatives that take precedence over social and environmental concerns. CEOs and literature meant for investors of companies such as Canada's Global Water Corporation, France's Lyonnaise des Eaux and Wetco, a water exporting company from Alaska, have often advertised their services as water sellers declaring that 'water has moved from being an endless commodity that may be taken for granted to a rationed necessity that may be taken by force'.

The neoliberal economic model strikes at the concepts of sanctity (as in Hindu culture) or trusteeship (as in Islam) and puts everything on sale like seeds and genes, culture and heritage, water and food. Perhaps air too would be commodified if scientific developments could bring in ways to ration it in future.

Even the so-called public–private partnerships are of no worth as private companies have to assure a return to shareholders even as they secure guarantees from the governments who enter into such partnerships. For instance, Chile had to guarantee a profit margin of 33% to Lyonnaise des Eaux as a condition of the World Bank, regardless of performance. What is obvious is that these monopolies usurp huge infrastructures raised by governments in the past, but are committed to their shareholders. Rising of water tariffs to an exorbitant extent by Bechtel brought the people in Bolivia on to the streets in the past leading to the government scrapping such partnerships ultimately.[16]

There are savants who view water as a metaphor, a metaphor for food and water. Others think of it as a symbol/metaphor for all the array of resources on which man depends for his survival. Some consider water as a 'social good' in the same league as air and food.

There are sociologists, anthropologists and historians who have undertaken studies of water with reference to and/or in relation to their area of study. There are scholars who view the problems associated with water from a legal point of view. There are those who view water from the vantage point of their religious books and dogmas.

Thus, we come across an astonishingly large variety of approaches towards the study of water. Each one of these studies is of considerable use as also of keen interest to persons working in a particular field. Some of these studies may be of interest to general readers. Our study here looks at how religious scriptures—in this case Quran—deal with the question of water. This study is focused on a specific area. Even so, the literature available in this niche is remarkably vast. In fact, the import being we are, as it were, touching only the fringe of the vast reservoir!

While summarizing the study, we could appreciate the import of what Quran affirms: 'there are enough resources to satisfy the legitimate needs of all human beings on the earth'. However, along with food water is not adequately available to

---

[16] http://democracyctr.org/bolivia/investigations/bolivia-investigations-the-water-revolt/bolivia%E2%80%99s-war-over-water.

the poor/weak in the society. Quran is particularly concerned that men (a) develop water resources; (b) use it equitably and judiciously without wasting it. They should avoid polluting the supplies along with their environment. Thus, the overwhelming emphasis in the Quran is on distributive justice.

Websites

http://english.islammessage.com/articleDetails.aspx?articleid=305

For zamzam: http://www.irfi.org/articles4/articles_5001_6000/abe%20zamzam.htm

For water clock: http://en.wikipedia.org/wiki/jayrun_waterclock

For Qanat system: http://www.academia.edu/432605/Technology_and_religion. The Qanat underground irrigation system.

For Ush'r on agricultural produce: www.muslimtents.com/shaufi/b

Perspectives on Water and the Bible

Y. Moses

Introduction

The world is faced with the hard reality of water crisis. There is, on the one hand, a fast depletion of water resources with major rivers drying up and, on the other, an increasing demand for scarce sources of water, leading to potential global water wars. There is an added burden on governments the world over to supply clean drinking water to their populations at affordable costs. In such a context, water conservation through a proper water management assumes greater significance and urgency. In terms of concerted efforts by individuals, communities, governments and global bodies. It involves willingness to identify root causes and a greater determination to rectify mistakes. Furthermore, it involves not only the adoption of appropriate techniques and technologies, but also the promotion of suitable values and attitudes. Fortunately, humanity has the advantage, especially in a globalized world, to learn from one another's experiences as also from one another's civilizational practices (Fig. 1).

The depletion and contamination of water resources and by extension the environmental and ecological imbalances caused by anthropogenic activity is a reflection of human deficiency either in the form of ignorance about nature's limitations or a sheer arrogant intention to dominate and exploit the natural world. If this situation were to be reversed, that is, if a further depletion has to be stalled and nature restored and preserved for generations to come, a review of social values vis-a-vis nature is absolutely essential. In this connection, looking into religions and looking back at pre-modern civilizations undergirded by multi-religious outlooks can help serve the purpose of drawing valuable insights for guidance in the present context.

Y. Moses (✉)
National Law School of India University, Bengaluru, India

Diocesan Training Centre, Anglican Church, Mauritius
e-mail: mosesy@hotmail.com

K.V. Raju, S. Manasi (eds.), *Water and Scriptures*,
DOI 10.1007/978-3-319-50562-6_3

Fig. 1 The bleak future for water. http://im.rediff.com/money/2009/dec/11water1.jpg

Religions retain a much needed sense of mystery about life and nature as also the capacity to animate people's minds and hearts towards acting with a great sense of responsibility. Christianity is one such religion.

Purpose

The purpose of this chapter is to draw from Christianity, one of the world's major religions, insights that can help promote conservation and proper water management. Here the focus is limited to identifying key Christian perspectives on water and their underlying values, starting from the Christian scriptures: the Bible. Identifying such values and principles can be critical to an evaluation of our own water-related practices and policies. The methodology involves extracting passages and verses from the Bible that refer to water, interpreting them in terms of their social and historical contexts besides relating their wisdom to the present water crisis. It describes water associated religious and social practices, while presenting some details about water supply and water management systems of civilizations during the biblical times. The information regarding the techniques and technologies employed in such distant pasts may not be of great use to our present complex contexts, but may serve an indicative purpose. While the focus is on biblical times, the chapter also highlights and reviews Christian doctrines and theological frameworks that could impact societal attitudes towards nature and its resources.

Bible and the Biblical Times

What the Christians regard as their Bible is a book of many books in two parts, the Old Testament and the New Testament. The New Testament, consisting of 27 books, reflects upon the life and teachings of Jesus Christ,[1] besides presenting an account

[1] Jesus Christ, upon whom Christianity is founded, was himself a Jew and his first followers were Jews.

of the early spread of Christianity. The 39 books of the Old Testament—ordered differently into 24 books—constitute the Holy Scripture to the Jews.

The books of the Bible contain different literary forms. They include poetry, wisdom, literature, oratory, letters and narrative history; figurative speech such as parables, similes, myths, metaphors, analogies and allegories. Throughout history, the Bible has been interpreted in a variety of ways—literally, morally and spiritually. What is significant for our purpose is the possibility of reinterpreting its message to address our own contexts and issues. This was indeed what had happened in the case of biblical writers themselves. For instance, the New Testament writers had reinterpreted the books and message of the Old Testament in the light of the person and work of Jesus Christ. In contrast to the assumption that the intention of the Bible is to give us accurate information, the biblical writers and readers both Jews and Christians up until the modern times did not think that the Bible had given a single, orthodox message and, therefore, were engaged in a constant reinterpretation. The principle applied through the ages is one of moving forward. The Jewish Rabbis and Christian Church Fathers used the old scriptures not for retreating into the past, but for propelling them into the uncertainties of the future. According to them, because the word of God contained in the Bible was infinite, a text proved its divine origin by being productive of fresh meaning.[2]

The period of biblical writings ranges from the eighth century BCE to the second century CE. Although the setting of the narratives in the Old Testament starts with the period beyond 1200 BCE, these narratives reflect the condition of eighth, seventh and sixth centuries BCE.[3] At these times, 'Israel'[4] was passing through a turbulent period with dramatic social, economic and political changes as they struggled for a settled national identity often in conflict with neighbouring imperial powers and civilizations in the Ancient Near East. The Old Testament writers were focused on human history, and this is important to note in understanding the Bible. They paid little attention to the cosmological myths that fascinated their neighbours—the Syrians and Mesopotamians in the Middle East. Apart from this distinction, Hebrew religion is comparable with the religion of their neighbours. The Israelites worshipped many gods even as others did. However, it was only since the sixth century BCE that monotheism came to be the main characteristic feature of their religion.[5]

[2] Karen Armstrong, *The Case for God*, p. 84.

[3] Ibid p. 38.

[4] A confederation of tribes collectively known as 'Israel' or the 'Hebrews' were bound together through a covenantal treaty. Their covenantal relationship was based on the principles of equality, freedom and social justice. All their laws and regulations reflected these values.

[5] Ibid, p. 40.

Fig. 2 Jesus turns water into wine. http://1.bp.blogspot.com/-x-u3mTKAIY0/VppFEB42AsI/AAAAAAAAByc/xrxRNu3hJfg/s320/alotin.vn_1404274370_huongptp2012122485938835_7.jpg

Biblical Perspectives on Water

The Bible has over 700 references to water.[6] Water in the Bible is used literally and figuratively sometimes as the setting and at other times as the element used by God to bless his people. The first miracle Jesus performed is with water, turning it into wine to cover insufficiency at a wedding. In so doing, he blesses the couple, saves the hosts from embarrassment and satisfies the guests (Fig. 2).

Significance of Water According to the Bible

Water in the Bible is regarded a blessing. In Isaiah 44, Israel is promised blessing that will come to them from God.

> *'But now hear, O Jacob my servant, Israel whom I have chosen! Thus says the Lord who made you, who formed you from the womb and will help you: Fear not, O Jacob my servant, Jesh'urun whom I have chosen. For I will pour water on the thirsty land, and streams on the dry ground: I will pour my Spirit upon your descendants, and my blessing on your offspring'.* ***Isaiah 44:1-3***

[6] www.quality-drinking-water.com/bible_and_water.html

Fig. 3 Parting of the Red Sea. https://www.google.co.in/imgres?imgurl=https://i.ytimg.com/vi/YnQxDjAVtJ8/hqdefault.jpg%3Fcustom%3Dtrue%26w%3D120%26h%3D90%26jpg444%3Dtrue%26jpgq%3D90%26sp%3D68%26sigh%3DG9nIAKHoA2a4KIUzEgmrq8iMfAk&imgrefurl=https://www.youtube.com/watch?v=mMOlfUo3JVY&h=90&w=120&tbnid=6h903mp6Dx76pM&tbnh=72&tbnw=96&usg=__bo3iAXcWvOv1bMIgKzScEn1qP-g=&hl=en-IN&docid=lSbFOuw-GJ-lvM

God is attributed with goodness because he gives rain and water. Eliphaz a friend of Job[7] declares God must be good and just because (Fig. 3):

He gives rain upon the earth and sends waters upon the fields ***Job 5:10***

The Bible recognizes the importance of water for thirsty people and faith in God's providence in spite of all odds. In the desert, the Israelites were thirsty and needed water. They needed enough for about one million people as well as all the livestock.

'And the LORD said to Moses,[8] *"Pass on before the people, taking with you some of the elders of Israel and take in your hand the rod with which you struck the Nile, and go. Behold, I will stand before you there on the rock at Horeb; and you shall strike the rock, and*

---

[7] The book of Job probes the depths of faith in the midst of suffering. It contains the folktale of a saintly man, Job who endures suffering even while being righteous. In the poetic language of the book, God is at work in the universe, even to bring rain on a land where no one lives. God reveals himself personally to Job and shares with him the vision of cosmic responsibilities.—Bruce M. Metzer & Roland E Murphy, ed. *The New Annotated Bible,* p. 625.

[8] Moses is regarded as the founder of Hebrew religion. The one who liberated the people of Israel from bondage in Egypt and led them in the desert for 40 years when he was instrumental in establishing a covenant (mutual agreement) between the Hebrews and their God, Yahweh based on the Ten Commandments. This incident of striking the rock for water happened at Mount Horeb, and it was believed that the supernatural rock followed the people of Israel through the rest of the pilgrimage to provide them water.

water will come out of it, that the people may drink". So Moses did so, in the sight of the elders of Israel'. ***Exodus 17:5-6***

In the same book of ***Exodus,*** water is pictured as a rescuing force. The infant, who would have otherwise been killed by Pharaoh's men, was rescued from the river Nile by Pharaoh's daughter. She named him Moses, *'because I drew him out of the water'.* ***Exodus 2:10***

Similarly, God parted the Red Sea to rescue the Israelites from the army of Pharaoh. With this act, God delivered and judged at the same time. The people of Israel were saved from being pushed back into slavery and the Egyptian army that chased them was drowned.

Even Noah[9] and his family were kept safe and carried by water until the great floods sent to destroy the wicked and the corrupt receded (Fig. 4).

'And all flesh died that moved upon the earth, birds, cattle, beasts, all swarming creatures that swarm upon the earth, and every man; everything on the dry land in whose nostrils was the breath of life died. He blotted out every living thing that was upon the face of the ground, man and animals and the creeping things and birds of the air; they were blotted out from the earth. Only Noah was left, and those that were with him in the ark'. ***Genesis 7:21-23***

Biblical writers see water as the source of life. That is why in the creation stories found in Genesis 1 & 2,[10] water pre-existed all other forms of nature and was from water that swarms of living creatures had emerged. According to the Bible, God created the world in 6 days. And on the 7th day, he rested. On day five of creation, God said, *'Let the waters bring forth swarms of living creatures...'* ***Genesis 1:20***

Water Covenants[11]

Because water was so precious in the Near-East, often covenants were made over wells or other sources of water. Also, wives for Isaac, Jacob and Moses were found at wells.

---

[9] The story of Noah and his Ark is the Great deluge or the flood myth found in the book of Genesis Chapters 6–9. The narrative indicates that the God of Israel intended to return the earth to its pre-creation state of watery chaos by flooding the earth for 370 days because of the world's evil doings.

[10] There are two creation stories in the book of Genesis, Genesis 1:1-2:3 and Genesis 2:4-24. The first was written in the sixth century BCE and the second in the eighth century BCE.

Almost every culture has developed myths of a lost paradise, from which men and women were ejected at the beginning of time. This was the case in the Ancient Near East, in Mesopotamia as well.

In the ancient Middle East, creation was regularly associated with temple building, and this Genesis myth was closely related to the temple built by King Solomon (c.970–30 BCE) in Jerusalem. Solomon's temple was apparently designed as a replica of Eden, once in Eden, Yahweh dwelt in the temple among his people. The temple was therefore a haven of *shalom.*

[11] Covenant is an agreement between two or more persons. The Bible mentions covenants made between and among people and covenants made between God and people. In the case of the latter

Fig. 4 Angel makes covenant with Hagar at a spring of water. https://tracygrierministries.files.wordpress.com/2011/04/hagar-and-ishmael1.jpg

The Lord makes a covenant with Hagar.[12] The Angel of the LORD found her by a spring of water in the wilderness, the spring on the way to Shur… The angel of the LORD also said to her, 'I will surely multiply your offspring so that they cannot be numbered for multitude… Behold, you are pregnant and shall bear a son. You shall call his name Ishmael, because the LORD has listened to your affliction. He shall be a wild donkey of a man, his hand against everyone and everyone's hand against him, and he shall dwell over against all his kinsmen'. ***Genesis 16:7***

Abraham and Abimelech make a covenant at Beersheba.[13] So they made a covenant at Beersheba. Then Abimelech and Phicol, the commander of his army, rose up and returned to the land of the Philistines. ***Genesis 21:32***

God makes a covenant with the Israelites at the spring He makes sweet. And he (Moses) cried to the LORD, and the LORD showed him a log, and he threw it into the water, and the water became sweet. There the LORD made for them a statute and a rule, and there He tested them, saying,

'If you will diligently hearken to the voice of the LORD your God, and do that which is right in his eyes, and give heed to the commandments and keep all his statutes, I will put none of

---

type, God always remained on a higher level and he makes the covenant as an act of grace. It involves promise by God and obedience on the part of the humans. In the former case, the contracting parties are on the same level.

[12] Hagar appears in the patriarchal tradition of the Jews. According to this tradition, the Israelites were Abraham's descendants, and he was considered as the Father of the Jewish nation. Sarah, Abraham's wife gave her maidservant to Abraham, as she herself was beyond the age of bearing children. Hagar bore Ishmael, after which quarrels ensued between the mistress and the maid leading to the latter being thrown out of the household. According to both Jewish and Muslim traditions, Ishmael was the ancestor of Arabs.

[13] There was a dispute between Abraham's servants and the servants of Abimelech, king of Gerar over certain wells which the former had dug. Abraham and Abimelech succeeded in reaching an agreement and swore to abide by it.

Fig. 5 God keeps mighty waters under control. https://sameapk.com/ocean-wave-wallpaper/

the diseases upon you which I put upon the Egyptians, for I am the LORD, your healer'. ***Exodus 15:26***

While the Bible appreciates the importance of water, it also sees the mighty waters, the seas and the oceans as dangerous and destructive, often symbolizing power that overwhelms humanity. However, God is more powerful and keeps these mighty waters under control (Fig. 5).

Deliverance from enemies is like being delivered from the deep ocean waters.

'For the waves of death encompassed me, the torrents of perdition assailed me, the snares of death confronted me. In my distress I called upon the Lord; to my God I called. From his temple he heard my voice, and my cry came to his ears'. ***2 Samuel 22:5-7***

In the New Testament, Jesus calms the storm. ***Mathew 8.25-27.*** Jesus and Peter also walk on water.

The monstrous seas or rivers, in contrast to the waters of Shiloh[14] that flow gently, symbolize the imperial powers and oppressive rulers who overtake Judah, allowed by God because of people's iniquities and corruption.

'Therefore, behold, the Lord is bringing up against them the waters of the River, mighty and many, the king of Assyria and all his glory; and it will rise over all its channels and go over all its banks; and it will sweep on into Judah, it will overflow and pass on, reaching even to the neck; and its outspread wings will fill the breadth of your land, O Imman'u-el'. ***Isaiah 8.7-8***

Spiritual Symbol and Metaphor

In the Bible, the word of God and water are considered important for life. The Bible compares the word of God with 'rain'. The believer is called to accept the word and water, cherish, respect and honour them both as the gifts of God upon freely

[14] The 'waters of Shiloh' refers to the canal which conducted the water from the spring of Gihon to the pool of Siloam in Jerusalem. The figure of speech is an effective one in contrasting Jerusalem's tiny stream with the mighty Euphrates, and at the same, suggesting the quietness and confidence which Ahaz, the king of Judah had rejected in favour of material power.—*The Interpreter's Bible, Vol. 5, p. 224.*

Fig. 6 River signifies living water. https://www.pexels.com/photo/nature-brook-creek-stream-4793/

bestowed. Like water is drawn from wells, so also the believers are advised to draw from the well of salvation which is God's word. The word of God and water sustain both physical and spiritual lives. A lack of water is often equated with spiritual dryness. Water becomes a symbol for the outpouring of God's spirit and blessing. God 'will pour water on the thirsty land and streams on the dry ground', ***Isaiah 44, 3*** (Fig. 6).

Living water is often spoken of in Judaic law.[15] Living water means river or spring water, as opposed to stagnant lake water. Purification rituals generally require living water. In the New Testament, the Spirit of God is also spoken of as living water, both quenching the thirst of and flowing from the believer. In the New Testament, Jesus is considered the source of living water. He says,

> *'But whoever drinks of the water that I shall give him shall never thirst; the water that I shall give him will become in him a spring of water welling up to eternal life'.* ***John 4.14***

Jesus made this offer in the context of the festival of booths,[16] which was observed as a reminder of water that flowed from the rock in the desert ***Num 20.2-13.*** It was observed annually for 7 days, during which time water was carried in a golden pitcher from the Pool of Siloam to the temple in Jerusalem.

The Bible uses water as one of the symbols for God's generosity. God is praised as the good shepherd who leads one to quiet waters ***Ps. 23, 2***. When Hagar and her son Ishmael ran out of water in the desert, God opened her eyes to see the saving

---

[15] Judaic Law refers to a set of Jewish rules and regulations that affect every aspect of life. They include those derived from the first five books of the Bible, known as Torah, those instituted by the Jewish Rabbis and those drawn from Jewish customs. It was believed that observance of such laws would turn the most trivial mundane acts, such as eating and getting dressed into acts of religious significance.

[16] Festival of booths also known as the festival of Tabernacles was an annual harvest festival of the Jews. The Israelites were required to form booths of branches of trees, palm leaves, etc. and dwell in them for 7 days during the festival. It brought back memories of their deliverance from Egypt, when they sojourned in the wilderness without a permanent habitat.

well ***Gen. 21, 15ff***. God sends rain on the righteous and on the unrighteous, just as God makes the sun rise on the evil and the good ones ***Matth. 5, 45.***

Furthermore, the Bible calls on people to emulate such generosity. God's generosity is to be reflected in the relationships between various human communities.

Water as a Healing Element

The importance of water is further heightened with its qualities of healing and cleansing. The Bible narrates several incidents where water was used to heal the sick and diseased. To give just two examples, a commander of the Army of Aram, Naaman, was healed of leprosy when he followed the instruction of Elisha, the prophet in Israel by way of dipping himself seven times in the River Jordan. ***2Kings 5.1-13.*** The book of Leviticus in the Bible which contains various laws has the following instruction for being healed of leprosy (Fig. 7).

> *'And he who is to be cleansed shall wash his clothes, and shave off all his hair, and bathe himself in water, and he shall be clean. After that he shall come into the camp, but shall dwell outside his tent seven days. And on the seventh day he shall shave all his hair off his head; he shall shave off his beard and his eyebrows, all his hair. Then he shall wash his clothes, and bathe his body in water, and he shall be clean'.* ***Leviticus 14:8-9***

In the New Testament, Jesus sends a man born blind to the pool of Siloam to wash himself to be healed of his blindness. The blind man returns healed. ***John 9.1-7.*** The pool of Bethesda in Jerusalem was considered to acquire healing powers whenever an angel entered it and stirred up the waters. A multitude of sick people, blind, lame and paralyzed would wait for their turn to enter the waters first to be healed at such times. ***John 5.1-4*** (Fig. 8).

Fig. 7 Naaman healed of leprosy. http://pearlsofpromiseministries.com/wp-content/uploads/2015/01/naaman-the-leper-washing-himself-in-the-jordan-x7-times.jpg

Fig. 8 The pool of Siloam. http://1.bp.blogspot.com/-AfjZKMZoONY/UmKSL7ZbL9I/AAAAAAAABIQ/6N8lIvZzHkQ/s1600/pool+of+siloam.JPG

Origin of Water

The Bible contains mythological anecdotes which use water to depict the conflict between good and evil, between God and Satan, the Leviathan.

A primordial battle took place between two gods in pre-creation. The victorious God tore the body of the dragon of chaos (associated with water) into two and separated the two halves with a firmament.[17] ***Genesis 1:6-7; Psalm 104:7-9.*** Our world sits between the waters above: the heavenly ocean, ***Psalm 104:3; 148:4*** and the waters below: the Deep ***Exodus 20:4; Psalm 136:6***. God will punish the Leviathan that dwells in the sea. ***Isaiah 27:1.*** The Leviathan or serpent is identified with Satan ***Revelation 12:9***; ***Job 3:8; 9:13; 26:12; 41:1; Psalm 74:14; 89:10; 104:26; Isaiah 27:1.*** God conquers the Ocean and the monster in it. ***Job 26.5-14***.

Ritual Cleansing and Purification

In the ceremonial system, washing was a permanent feature. Priests were washed at their consecration ***Exodus 29.4.*** Levites[18] were sprinkled with water ***Numbers 8.7.*** Special ablutions were demanded of the Chief Priest on the Day of

[17] It is difficult to understand the creation story in Genesis without a reference to the Mesopotamian creation hymn known from its opening words as the *Enuma Elish*. This poem begins by describing the evolution of gods from primordial sacred matter and their subsequent creation of heaven and earth, but it is also a meditation on Mesopotamia.

[18] Levites, the descendants of Levi Tribe of Israel, worked as assistants to the priests in the Jewish Temple.

Atonement[19] ***Leviticus 16.4, 24, 26,*** of the Priest in the 'water of separation' ritual[20] ***Numbers 29.1-10,*** and of all men for the removal of ceremonial defilement ***Leviticus 11.40; 15.5; 17.15; 22.6; Deuteronomy 23.11.*** The laver before the tabernacle was a constant reminder of the need for cleansing in the approach to God ***Exodus 30.18-21.*** A developed form of this ritual ablution was practised by the Qumran sect[21] as also by a variety of Jewish Baptist sects which flourished before and after the turn of the Christian era. These provide a background to John's baptism of repentance and to the Christian baptism of cleansing, initiation and incorporation into Christ.

Use of Water in Baptism

Christian baptism symbolizes the conversion of the baptized person as a new person, having decided to bury his sinful past and lead a new life under the guidance of God's Holy Spirit. The act involves either immersion in deep water or sprinkling of water. The immersion additionally indicates sharing in the death and resurrection of Jesus Christ. Urban churches and others that have access to water, and those who still wish to maintain the Jordan River baptism motif, have their churches designed in such a way that there is a pool at the altar that can be opened and filled with water during baptism services. Others carry out their baptisms in swimming pools, while a majority of mainstream churches are content to carry out baptism by effusion (sprinkling) (Fig. 9).

Because of its importance in biblical times, water came to be used in Christian religious ceremonies as a sign of cleansing. Healing, baptism, and foot-washing[22] symbolize religious uses of water. Water, through God's Spirit, transforms or changes the lives of those who receive it. Increasingly, water or oil is specially anointed and used to heal the sick. People who go on pilgrimages to the Holy land and special Shrines bring with them the anointed water. Thus, from the beginning to

[19] It is the Jewish annual day for atoning for the sins of the whole nation including that of the Chief Priest and the priests of the Holy Sanctuary. It is observed on the 10th day of the 7th month by abstinence from daily labour, by holy convocation and by fasting.—*The New Westminister Dictionary of the Bible,* p. 78.

[20] Water of separation means water for the removal of impurity. This special water (ashes of a heifer mingled with living water) is sprinkled on the 3rd and 7th day on all those defiled through contact with a corpse.

[21] A separatist Jewish community that held esoteric doctrines and practices concerning maintenance of strict holiness and communion with Angels.

[22] The root of this practice appears to be found in the hospitality customs of ancient civilizations. The host would provide water for guests to wash their feet, provide a servant to wash the feet of the guests or even serve the guests by washing their feet. This is mentioned in several places in the Old Testament of the Bible as well as other religious and historical documents. Foot washing was a sign of humility. Following the example of Jesus, several denominations observe foot washing in the churches on Thursday before Good Friday.

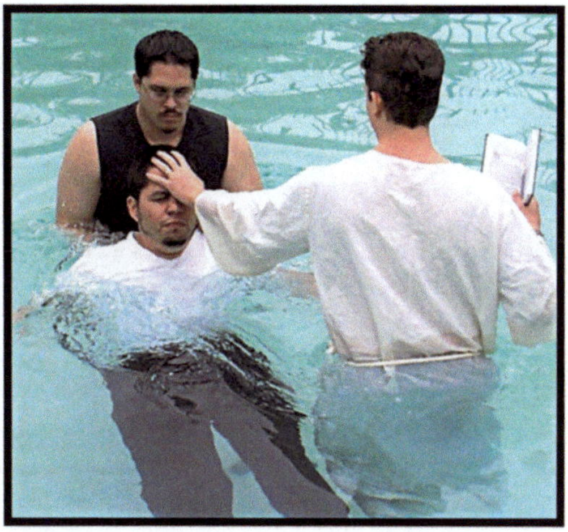

Fig. 9 Immersion Baptism. http://a339062bb12acaf0447f-fe54b552272646221985f1a127513c68.r76.cf2.rackcdn.com/B0F58A23-9523-47E6-85EB-88A1624224D5.jpg

the end, water flows through the pages of Bible. From the rivers that flow from Eden to Ezekiel's[23] river of living water and to the river flowing from the throne of God in John's Revelation[24]; rivers are signs of God's provision, blessing, healing and overall life-giving.

Water Festival

The Jewish water festival called the *sukkot* festival was celebrated annually during the festival of Tabernacles (Fig. 10).

In the mornings, on each of the 7 days of the feast of tabernacles, the priest would go to the pool of Siloah and fill a gold flask with water. They would go back into the city and pour the water simultaneously with wine from another bowl. This procession was accompanied and followed by music, dance and general celebrations. This festival was perhaps based on of a biblical verse found in ***Isiah 12:3,*** *'With joy you will draw water from the wells of salvation'*. It also might be related to folk traditions of offering water at the beginning of the rainy season, hoping for abundant rains.[25]

[23] Ezekiel was a priest and prophet to the people of Israel while they were in exile in Babylon in the sixth century BCE. His work marks the transition from pre-exilic Israelite religion to post-exilic Judaism. Judaism enabled the people to believe that their God resides amidst his people wherever they are, even in exile and not confined to his Temple in Jerusalem .Ezekiel's innovative use of vision laid the foundation for symbolic universe of apocalypticism.

[24] The book of revelation found as the last book in the New Testament, supposedly written by John, a disciple of Jesus Christ, belongs to the apocalyptic literature that envisions the end times.

[25] http://www.jewishencyclopedia.com/articles/14,794-water-drawing-feast-of

Fig. 10 Jewish water festival. https://kumi07.files.wordpress.com/2010/09/priest-of-tabernacles.jpg?w=1272

Fig. 11 Water Cistern. http://www.hydriaproject.net/images/cases/cyprus_Larnaka/T3_01_round_wells.jpg

Technologies in Water Management

Awareness Regarding Water Cycle and Hydrology in the Bible

Coming to science and technology related to water during the biblical times, it could be said that there was, in ancient times among the people of the Bible, some basic knowledge about water cycle and hydrology (Fig. 11).

> *'For as the rain and the snow come down from heaven, and return not thither but water the earth, making it bring forth and sprout giving seed to the sower and bread to the eater, so shall my word be...'* ***Isaiah 55:10-11a***

> *'All streams run to the sea, but the sea is not full; to the place where the streams flow, there they flow again'.* ***Ecclesiastes 1:7***

'Who calls for the waters of the sea, and pours them out upon the surface of the earth-the Lord is his name'. ***Amos 9:6b***

'For he draws up the drops of water, he distils his mist in rain which the skies pour down, and drop upon man abundantly'. ***Job 36:27-28***

Traditional Sustainable Practices and Water Storage, Conservation and Supply Systems in the Biblical Times

In the terrain where they lived, the people in Palestine during the biblical times, depended on discovering, digging, fighting for, and maintaining wells and springs rather than rivers. Natural springs determined the location of settlements.

Water storage made all the difference to survival during the dry season. They stored water in cisterns, public pools and aqueducts, which were sometimes very extensive, bringing water to cities or between pools.

Cisterns are usually pear shaped with a small opening at the top which can be sealed to prevent accidents and unauthorized use ***Exodus 21.33, 34***. Most homes in Jerusalem had private cisterns, ***2 Kings 18.31, Proverbs 5.15*** but there were also huge public cisterns: one in the temple area having a capacity for over two million gallons. By 1200 BC, cisterns were cemented, thus permitting large settlements in the barren Negeb region ***2 Chronicles 26.10.***[26]

Artificial pools were dug inside walled cities and often fed through a tunnel leading from a spring outside, ensuring supplies in times of siege. Examples have been found at Gezer, Megiddo, Gibeon ***2 Sam 2.13*** and elsewhere. 'Hezekiah's tunnel' and the pool of Siloam lie at the southern end of Ophel at Jerusalem ***John. 9.7, 11; Nehemiah 3.15***. The pool of Bethesda was usually located in the north-eastern corner of the city, near the sheep Gate. During the summer, water collected in pools during the winter and spring, formed an important source of supply.

Water Supply and Water Management Systems in Civilizations of Biblical Times

In Mesopotamia[27]

Two sources of drinking water in Mesopotamia were the Twin Rivers and their canals. For many cities, these remained the main sources of water down through the first millennium BCE. But some palaces, especially in Assyria, got their water supply from deep wells, free of pollution. For the most part, large cities were built near

[26] J.D. Douglas, *The New Bible Dictionary ed. The New Bible Dictionary,* p. 234.

[27] Taken from Karen Rhea Nemet-Nejat, *Daily Life in Ancient Mesopotami,* p. 110-11, 253–55.

Fig. 12 Sennacherib's Acqueduct in Jerwan. https://www.amazon.co.uk/Sennacherib-Aqueduct-Nineveh-Assyrian-Mountain/dp/B0087V0478

water sources. Smaller cities survived if they had many springs, wells, aqueducts or cisterns (Fig. 12).

In Nimrud, many wells were dug to a depth of 90 ft to protect the city's water supply in times of siege. In 1952, excavators while clearing one well, found it still able to provide 5000 gal/day. In that well, archaeologists found a wooden pulley wheel bearing the wear of rope marks, and many pots, some with ropes still around their necks forming a chain of vessels operated by a windlass to draw water from the well.

King Sennacherib had an aqueduct constructed in Jerwan, a village 9.6 km away, to supply Nineveh, his capital, with water for drinking and irrigation. This aqueduct, over 270 m long, anticipated later architectural plans in every detail. The entire structure used about two million stones, each weighing a quarter of a ton. The water flowed over the aqueduct floor which was hardened earth waterproofed with bitumen and lined with stone. The aqueduct extended over the valleys to arches and was fed by many small streams thereby guaranteeing an adequate supply of water to the city.

Agricultural land was best classified by its water supply, which regulated farming, the types of crops, the amount and dependability of yields, and the total area of land cultivated. Mesopotamia had two kinds of agriculture, dry farming in the north (Assyria) and irrigation farming in the south (Babylonia and Sumer) (Fig. 13).

Dry farming relied only on natural rainfall and was practised in northern Mesopotamia. Large-scale irrigation with complex canal systems supplemented by natural rainfall was used in southern Mesopotamia.

Irrigation was necessary for crops because the salinity of the soil was a problem early on. Water was channelled to fields from major water courses through branch canals and feeders, which often ran along the tops of artificial dikes. The width of the primary water courses could be 120 m or more; the branch canals were as narrow as 1–1.5 m in width and 0.5–2.25 m in depth, with a length just under 2 km. Weirs were built to raise the water level in the main streams, whereas Bunds were to protect against floods. Breaches were repaired with earth and bundles of reeds. Productive quay walls of baked bricks set in bitumen were built to guard against

Fig. 13 Irrigation farming in Babylonia. https://sites.google.com/a/student.sthelenaunified.org/rls-ancient-civilizations/_/rsrc/1403210451345/units-of-study/ancient-mesopotamia/mesopotamia/event-c-building-and-maintaining-a-complex-irrigation-system/Event%20C%20Euphrates%20River%20Canal.jpg?height=254&width=400

erosion across the canal banks at critical points. To prevent scour, pebbles or stones were used to cover canal beds (a practice recorded as early as 2400 BCE). The construction and maintenance of canal systems were considered an important duty (as well as an act of piety) to be executed by Mesopotamian kings, but cleaning and dredging accumulated silt in other canals was under the jurisdiction of the local authorities. The rivers and canals provided drinking water for people and animals, irrigated vegetation and created a cool, green world along their banks.

The waters of the Tigris and Euphrates came in spring when the fields were full of standing crops. A flood would have proved disastrous at this time, so a complicated irrigation system was set up. Strongly reinforced levees were built to keep the water in rivers. To obtain an efficient gravitational flow, the canal systems were long and well maintained.

Water was always in short supply in the growing season, as a great deal of water was lost through evaporation and seepage. The timing of watering was critical. The Mesopotamian farmer had to deal with the problem of salinity as irrigation water evaporated, slowly deteriorating the soil structure. A great deal of labour was necessary to restore fertility.

Irrigation required solving four problems: (1) supply (getting water to land that could be cultivated); (2) storage (keeping water where needed); (3) drainage (disposal of water when no longer needed) and (4) protection (keeping away unwanted water).

As far back as the third millennium BCE, gardens or small areas were watered by hand from wells or streams, but for cereal crops, the most practical system was gravity-flow irrigation. Water was brought to fields by successively smaller branching canals or by aligning major water channels parallel to the main rivers. Two controls were essential, outlets and regulators. Outlets or sluices could be as small as a hole opened in the side of a canal bank to divert water into a distributary channel.

Regulators were used for catching the entire main stream in order to raise water to the appropriate level. Temporary regulators were constructed from reeds, while the permanent installations were made of baked bricks and bitumen.

Water Supply in the Roman Empire[28]

Water was brought to houses by channels through walled conduits, lead pipes or earthenware pipes (Fig. 14).

Specifications for conduits: at least 6″ drop for every 100 ft to be covered by arches to avoid sun. A reservoir was made where the channel met the city walls. The reservoir was divided into three water holding tanks. One led to all city basins and fountains, one to the baths and one to private houses. Channels were made and covered above ground, or if they were to be tunnelled through hills, they were either cut through stone or constructed with soil or sand.

Lead pipes started with a reservoir at the source. Pipes ran from this reservoir to a reservoir inside the city. Pipes took advantage of the water pressure at their source to push water over low hills. Earthenware pipe systems were similar to lead pipes, fitted together and joined with quicklime and oil. They were easier to repair and gave healthier water. Romans were aware that lead had some negative effects, but still used lead for much of their water transport.

Both wells and cisterns were used as sources of water. Cisterns were made double or triple to allow sediment to settle down keeping water pure. The distribution

Fig. 14 Roman water pipes. http://affordablehousinginstitute.org/blogs/us/wp-content/uploads/roman_water_pipes.jpg

[28] Taken from Naphtali Lewis & Meyer Reinhold, ed. *Roman Civilization: Source Book II: The Empire,* p. 304–306.

and financing of the aqueducts and the other pieces of Rome's water system were described within Rome's legal system. Mostly they were handled locally, with power delegated by Rome to communities.

Summary of Biblical Perspectives

We have thus far seen, on the basis of biblical passages and verses, how biblical people had perceived the benefits and qualities of water, their understanding of water cycle and hydrology, their use of water in purification rituals and initiation practices such as baptism. We also have noted their celebration of water through festivals and water sustainable practices and management technologies. It is true that, in the Bible, water was for the most part used as a metaphor and symbol to convey insights about spirituality, morality and social ethics. They neither had the occasion nor the need to write a treatise about water and yet we could discern from their usage, the significance they had attached to it as a life-giving and life-sustaining natural resource. So much so, that they considered water as a gift and linked it to God's creating, blessing and saving work. While their science about the water cycle was basic, they clearly differentiated the value of running water from the potential dangers of stagnant lake water. For them, river water or spring water was the living water, which they generally required for their purification rituals. The importance of living water to them was furthermore heightened with its qualities of healing and cleansing.

Rulers' Far-Sighted Responses

As for sustainable practices, they adopted conservation methods best known to them and best suited for their geographical conditions. They did what was the only option available, which was to secure their water through collecting rainwater or digging wells and storing it in cisterns to overcome shortage in needy times. The Bible narrates stories of the biblical people ranging from their time as pastoral people to agriculturists and to city dwellers. To meet the water needs of their citizens, rulers of biblical times, like Hezekiah, the King of Judah, or Sennacherib, the King of Nineveh and others constructed and maintained canal systems, aqueducts and pools. They shouldered the responsibility of ensuring an adequate water supply to their subjects even during times of siege.

Going beyond a homiletic usage, the Bible also narrates incidents of controversies and conflicts over sharing of water resources. It would be helpful to take a closer look at the manner in which such conflicts were managed.

Fig. 15 Semi-nomadic people and their flocks fought over water. http://blog.joins.com/usr/a/la/alamin/1210/508a212e6eb0a.jpg

Philosophical Approaches to Conflict Resolution

Understandably, water in the biblical times, was a source of conflict, considering its scarcity. It is possible that some of the battles fought between neighbouring nations and empires were mainly for gaining access to and controlling the few rivers that existed, such as the Euphrates, Jordan and Nile. Early in history, people in Palestine fought over wells that they had dug. The availability of a well or cistern was of particular importance for semi-nomadic people and their flocks. As the conflict between Abraham and Abimelech shows, the ownership of a well could easily become the subject of quarrels between those with large flocks ***Gen. 21, 22ff; 26, 15ff***. Yet conflicts were resolved amicably with a philosophical approach (Fig. 15).

While Genesis tells of the struggles between ancient peoples over water, it also reflects God's will that water is for all, not for one particular people over others. Isaac moved from *Ezek* and *Sitnah,* the wells of dispute, to another place where he dug one more well which he called *Rehoboth ('broad space'),* a name that does not reflect his skills in finding water, nor his diplomatic or military abilities, but his recognition that land and water are the natural gifts of God: 'Now the Lord has made room for us, and we shall be fruitful in the land'. There was room for both Philistines and Israelites to flourish in the land; God provided water for both. Wherever and whenever there is conflict over water today, the biblical narrative reminds us all that water is God's gift, and that it is never anyone's property.

Recognition of the Rights of the Poor

When it comes to water rights, the Bible clearly comes heavily on those who deny the poor the right to water. To offer water to one, who is thirsty, even to the enemy, is a basic criterion for right relationships ***Gen. 24, 15ff; Prov. 25, 21; Matth. 25, 42;***

Fig. 16 Meeting the needs of the poor a priority. http://image.slidesharecdn.com/presentacinaguafinal-131203123706-phpapp01/95/concientizacion-del-agua-4-638.jpg?cb=1386074338

Rom 12, 20. Only a villain or a fool will deprive the thirsty of a drink. ***Isaiah 32, 6; Job 22, 7***.

Having to pay for water is considered a mark of oppression and unjust treatment. ***Num. 20, 19; Lam. 5, 4***. The writer of Matthew's gospel declares that the one who withholds water from 'one of these little ones' ***Matt 5:42*** is not righteous and will not experience God's blessing (Fig. 16).

The prophets in the Bible call for the practice of justice using water as a metaphor. Says Amos, 'Let justice roll down like waters, and righteousness like ever flowing stream'. ***Amos 5:24***. Justice to the poor was called for in the life of the people of Israel as part of their covenant (constitutional) obligation made at Mount Sinai after their liberation from bondage in Egypt. We must remember that the Old Testament, besides containing words of wisdom, is deeply rooted and reflects upon socio-political and historical situations of the people of Israel. It considers the people of Israel as a chosen nation, not because of any merit intrinsic to them, but because they were in need. They were poor and oppressed when God, through his righteousness, delivered them from bondage in Egypt and chose them as a witness to His care and concern for all the oppressed people of the world.

The Old Testament contains religious, moral and civil laws to guide Israel to build and sustain a just community. For the faithful Jew, these laws reflect the mind of God. The distinguishing feature of Israel's laws when compared with the laws of its neighbours is that, they were clearly more humanitarian. While the laws of the neighbouring communities put their emphasis on property and profits, Israel's law protected the poor. ***Deuteronomy 15.7-11*** demanded that the rich be generous to the poor. ***Leviticus 19.9-10*** gave the poor special gleaning rights. Even more interesting and important was the institution of a periodic reordering of economic life.

The prophetic tradition in the Old Testament started with the institution of monarchy in Israel. The prophets served as constructive critics of socio-economic and political trends, often under a great risk to their own lives. They repeatedly reminded the people of Israel about their covenant obligations to do the will of God, which, among other things, were to do justice to the poor. They condemned worship bereft of works of justice. And they associated injustice with sin. The prophet Daniel said, 'Stop sinning, do what is right and be merciful to the poor'. ***Dan 4.24-28.*** Jeremiah regarded a fair trial to the poor as equivalent to loving God ***Jeremiah 22.12-19.*** Isaiah was even more forthright. He says that when the rich and the poor stand opposed to each other, God is on the side of the poor ***Isaiah 1.17.***

'Justice to the poor' was also practised by Jesus Christ as an essential aspect of his mission in the world. And all his disciples then and now are called to analyse, assess and transform societies in the light of the cry of the poor for justice. Jesus based his teaching and patterned his ministry along the prophetic lines. Declaring his manifesto at the beginning of his ministry, he said:

> *The Spirit of the Lord is upon me, because he has anointed me to preach good news to the poor. He has sent me to proclaim release to the captives and recovering of sight to the blind, to set at liberty those who are oppressed, to proclaim the acceptable year of the Lord.* ***Luke 4:18-19***

Sure enough, Jesus offers liberation to the rich as well. He reminds the rich of the fact that true security lies not in their possessions, but in God, the Father who provides. Rather than seeking more wealth, Jesus calls on the rich to seek first the kingdom of God and its values. An important kingdom, value is justice—justice to the poor. According to Jesus, all those who do not respond to the basic needs of the poor have no place in the kingdom. The parable of the last judgement illustrates this point. It is an indictment against individuals and societies that do not respond to the cry of the poor for food, water and clothing (Fig. 17).

With such a value laden teaching and attitudes towards water as elicited thus far, it would be hard to believe that biblical teaching could have played a major role in

Fig. 17 Jesus calls the rich to respond to the basic needs of the poor. https://i.guim.co.uk/img/static/sys-images/Business/Pix/cartoon/2010/3/9/1268167648497/A-girl-carrying-water-in--001.jpg?w=470&q=55&auto=format&usm=12&fit=max&s=c23507580718bae41288a6ead739b597

contributing to the crisis of depletion and unfair distribution of earth's resources including water resources. The ecological and environmental crisis has been squarely placed at the door step of western civilization that pioneered and propagated an unlimited exploitation of earth's resources in the name of development. Part of the blame is put on some Christian doctrines which allegedly have reinforced an unrestrained pursuit of economic development goals. It has been assumed by some scholars that a secular scientific exploitation of natural resources, which began with the onset of modernity in the seventeenth century, was allegedly reinforced by the concept of dominion, associated with the creation story in the Bible.

Search for an Appropriate Theology and Ideology

The charge against Christianity's role in the ecological crisis was led by Lynn white Jr., who in his famous article, 'The Historical Roots of our Ecological Crisis', written in 1967, maintains that in destroying native religions, and their belief in God or gods dwelling in the world, Christianity had desacralized the natural world, thus laying it open to exploitation. He observes that the historical nature of Judeo-Christian revelation, with its linear view of time, was an important factor in the emergence of the doctrines of human progress in western society.

While referring to the creation story, which contains the command to humans to be fruitful, multiply and subdue the earth (***Genesis 2:15***), and the verses after the flood 'Every moving thing that lives shall be food for you; and as I gave you the green plants, I give you everything' (***Genesis 9:3***), White maintains that Judeo-Christianity preaches humans are separate from and superior to the rest of nature, which is there to be used and dominated by humankind. White observes that when such Christian attitudes were combined with the explosion of technological power, as applied science in the nineteenth century, the way was open for an apocalyptic ecological despoliation.

However, White's thesis has been countered by other scholars. Many of the repudiations focus on the possibility of interpreting the Judeo-Christian message in exactly the reverse of White. The Book of Job and Paul's Epistles to the Romans set humans apart from all other forms of life and matter because he says God has willed this role for them. Human beings are set here as stewards, responsible to their Creator for all they do with the world over which they are given dominion. Dominion, according to White's critics, could simply mean 'ruling' or 'governing' the world, not exploiting nature, come what may.

They argue that White's argument is 'partisan and overgeneralized' and that Genesis characterizes humans as life tenants of nature, not freeholders. There is much more evidence than is usually acknowledged for other 'more beneficent Christian attitudes to the environment and to non-human nature'. Christian teachings about nature have been diverse and contradictory, but they have not typically been exploitative.[29]

[29] R.L. Sarkar, *The Bible, Ecology and Environment,* p. 150.

Emergence of Theologies of Nature

Christianity is blamed partly because of the modern existential theologies that over emphasize interpersonal relationships to the neglect of communal relationships and man's relationship with nature. The lop-sidedness of such theologies stands corrected by new theologies of liberation and theologies of nature. The main contribution of liberation theologies lies in their methodology. They analyse all social crises from the perspective of the victims in modern societies and give a priority to poor peoples' demands and vision of alternative societies. This is a good reminder, while dealing with water-related problems since there is a tendency in a technocratic civilization to find technocratic solutions only.

In the recent decades, and with the realization of an ecological crisis, the Christian Church has begun to show interest in developing theologies of nature. These theologies caution us that in our attempts to produce more material goods to meet the increasing demands, we should not and cannot sacrifice nature. As responsible stewards, we are held accountable to God for what we do to nature. Therefore, they call for a balance between the instrumental value and the intrinsic value of nature.

Emergence of Social Ecology

There is no doubt that the economic growth model of development with its assumption that there are no limits to natural resources is the root cause underlying the ecological crisis we face today. This model and its free-market ideology should be critically reviewed in order to ensure eco-justice and justice to the poor. The alternative model to economic growth model is social ecology (Fig. 18).

Social ecology is distinguished from shallow environmentalism that blames the ecological crisis on inappropriate technology, overpopulation or industrial growth,

Fig. 18 Humans should live in harmony with nature. http://kamkankouken.jp/tourism/common/img/activity_list/masyu_wk.jpg

and ignores the fundamental cause deeply rooted in the institutionalization of domination and hierarchy and the authoritarian mentality sustaining it. Social ecology considers stewardship and justice as important principles. It recognizes the reality of social and political institutions of domination and hierarchy that block the liberation of nature and humanity and promotes a dynamic social movement to transform the way we find our place within nature and to change our relationships with each other and non-human world. The main concern of the social ecologists is to change society in order to change our relationship with nature. The social ecologists tend to be more humanist, believe in participatory democracy, and exhibit a greater sympathy for the poor and the oppressed. The recent Christian theologies of nature are patterned after the ideology of social ecology.

Lessons for the Future

As already stated in the introduction, the purpose of this study is to draw insights or lessons from the Bible and from water-related practices of people and civilizations influenced by biblical teachings. The idea is to use the derived lessons as parameters for addressing our own water-related problems and also for promoting policies that can sustain the availability of water resources for future generations. With that purpose in view, we had set out to find some guidance for water conservation and proper water management. We had also noted the scope and limitations of this study. As far as limitations are concerned, it has become clear that the technologies of the ancient civilizations were far less advanced than we have at present. However, the philosophical, social and ethical values and principles advocated and promoted in the past by the biblical patriarchs, prophets, priests and teachers still hold good in our individual and social life, particularly in relation to how we ought to treat nature. If we treat our natural resources as God given common goods, we would desist from an over exploitation of the same simply because it is profitable to do so. Also our mechanisms with respect to the distribution of the common resources will apply the principles of social justice and need-based priority. It is true that the biblical stories and narrative histories are anthropocentric, but with a gradual realization of the limitations of nature, biblical interpretations (theologies) should necessarily change in accordance with changing realities. Fortunately, this is happening in the Christian Church leading to changed perspectives, attitudes and approaches to water resources and their limitations. A significant contribution is being made by the churches the world over towards water conservation and water management, primarily through their educational programmes and processes. Here are three examples of initiatives taken by the Christian churches.

The changing Christian theological teaching about nature is taking a concrete shape in the Churches' responses to the crisis of water. The attempts made are primarily related to building awareness regarding the water crisis, and encouraging people, especially children and youth, to develop proper perspectives. Three good examples of such initiatives are: the global Ecumenical Water Net Work, the United Church of Canada and India Peace Centre of the National Council of Churches in India.

The Initiatives of Ecumenical Water Network

The Ecumenical Water Network is a global platform of churches and Christian organizations for promoting people's access to water around the world since its inception in 2006.

Based on the understanding that water is a gift of God for all creation and a fundamental human right, the Ecumenical Water Network (EWN) brings together churches, Christian organizations and individuals to promote universal access to water and sanitation, and the sustainable use and responsible management of water, so that all may live their lives with dignity and with respect for the integrity of creation.

EWN builds on the experiences of churches and ecumenical partners worldwide. During its annual Lent campaign, called The **Seven Weeks for Water**, it publishes biblical reflections, liturgical materials, background resources—and also ideas for what churches, congregations and individuals can do in their respective contexts.

EWN has also placed a special emphasis on involving and encouraging **young people**, for example, by organizing trainings like the 'Summer School on Water' and 'Youth for Eco-Justice' in cooperation with other departments of the World Council of Churches and the Lutheran World Federation. These youth initiatives have a ripple effect with many local initiatives emerging out of these interventions.

EWN has also highlighted water at key international church events, such as the general assemblies of the regional council of churches and they, in turn, have responded positively, by affirming key statements and declarations of the EWN on human rights to water and sanitation on many occasions.

The Initiatives of the United Church of Canada

The Justice, Global and Ecumenical Relations Unit of the United Church of Canada has developed a five-session congregational study entitled *Waters of Life*. The study kit contains two other resources—a daily devotional booklet and a lectionary-based resource for worshiping leaders—both of which use water as a predominating biblical touchstone and metaphor. It is used with children during Sunday school, as a special weeknight activity for the 6 weeks of Lent, or as a daylong Saturday or March break activity. It is designed especially for use in a Workshop Rotation Model, but could be adapted to graded classes or a multi-age one-room school model (Fig. 19).

The following principles form the ethical and theological grounding for the United Church of Canada's justice work on water issues:

- Water is a sacred gift that connects all life.
- Access to clean water is a basic human right.
- The value of the earth's freshwater to the common good takes priority over any possible commercial value.
- Freshwater is a shared legacy, a public trust, and a collective responsibility.

Fig. 19 Water need. http://www.usnews.com/dims4/USNEWS/34075f2/2147483647/resize/1200x%3E/quality/85/?url=%2Fcmsmedia%2F22%2Fe94078947a2c3a2f92e1485a9169bd%2F2020-4

The areas of focus for study of water with children are

- importance to the world
- human responsibility in terms of caring for creation (water)
- of biblical images of water
- water as a gift

India Peace Centre

A similar promotional work on water-related issues is done by the India Peace Centre situated in Nagpur, Maharashtra. Started in 1990 by the National Council of Churches in India, this centre has published the Children's Study Series on clusters of issues covering Ecology Environment, militarization-disarmament, human rights and justice-peace.

In addition to building awareness through educational programmes, Churches and related partners are involved in undertaking water projects. Church-sponsored projects cover a broad range of activities, like constructing and managing water systems, building hygienic toilets, or helping small-scale farmers make a more efficient and sustainable use of water for agriculture. Agencies like Norwegian Church Aid and Church World Service are implementing water and sanitation projects that aim at improving and protecting the health and livelihoods of communities in poor countries.

Conclusion: Contemporary Issues and the Way Forward

The introduction defined in a perfunctory manner our contemporary water crisis as an all-around depletion and degradation of water resources in the world besides indicating that there were dilemmas associated with these shortages. After having presented a fair account of the biblical perspectives on water and identifying their underlying values, as stewardship or responsibility, equality and social justice, I now wish to conclude the study, by identifying at least three different aspects of the water crisis and making a few suggestions by way of indicating a few areas for further action by individual citizens, neighbourhood communities, local governing bodies, state and national governments.

Three Aspects of Water Crisis

The water crisis in the world today involves three related problems of scarcity, pollution and water disasters

a. *The problem of water scarcity*
   The problem of water scarcity is borne by the fact that though the planet Earth is a living body with water constantly flowing through it, there is an increasing shortage of water being experienced with many rivers becoming dry.[30] In the recent decades, shallow groundwater has become an important source of water for irrigation, but has also led in most places to over-pumping of aquifers. Even drinking water is becoming scarce. Millions of people around the world struggle to secure safe water to meet their basic needs. In India, it is estimated that the national cost of women fetching water is 150 million women working days per year, equivalent to a national loss of income of ten billion rupees.[31]
b. *The problem of water pollution*
   Water pollution is visible everywhere. Our rivers are being polluted by the manufacturing and agricultural industries with municipalities adding to the problem through dumping of the sewage. An inappropriate use of fertilizers and pesticides is polluting drinking water, rivers and lakes. And water scarcity forces people into consuming contaminated water, leading to water-borne diseases, in the process. According to the United Nations Water Report, more than two million people in the developing countries, most of them children, die each year of diseases associated with unsafe drinking water, inadequate sanitation and poor hygiene (Fig. 20).[32]
c. *The problem of water-related disasters*
Water-related natural disasters such as floods, tropical storms and tsunamis are affecting millions of people. Between 1991 and 2000, over 665,000 people died in 2557 natural disasters, of which 90% were water-related events such as floods, typhoons and hurricanes.[33] Such disasters are a direct effect of climatic

[30] The Nile River doesn't reach its end, the Colorado River, the Yellow River in China for the most part don't flow any more to the sea.

[31] www.un.org/waterforlifedecade

[32] Ibid.

[33] Ibid.

Fig. 20 A polluted water way in Ipswich. http://media.mnn.com/assets/images/2009/06/pollutedwater.jpg.653x0_q80_crop-smart.jpg

changes that humans have caused through an excessive deforestation. In addition, human activity is affecting water ecosystems. For instance, water used as a coolant in generating stations is poured back into the rivers, which raises the temperature and speeds up certain biological processes, imposing sharp changes on aquatic life.

The world community is indeed concerned with the water crisis and all its related problems. Hence, both individual governments and the United Nations have stepped in to address the problems. Two important World Conferences, The Millennium Summit in 2000 and the Johannesburg Summit in 2002 are significant in this regard. The Millennium Summit issued a millennium declaration emphasizing the need for all countries to stop the unsustainable exploitation of water resources. Governments addressed this issue at the Johannesburg Summit, also called the World Summit, on Sustainable Development. They agreed to develop integrated water resource management and water efficiency plans and committed themselves to reducing by half the proportion of people without access to safe drinking water by 2015, the year to mark the end of *International Decade for Action: 'Water for* Life'. While the concern and efforts of the UN and the governments have been appreciated though belatedly, the promotion of privatization of water by the World Bank and the International Monetary Fund through Multi-National Corporations has come under a severe criticism and resistance in several countries. There is this dilemma of having to provide for safe drinking water on the one hand and, on the other, making it available for all at affordable costs (Fig. 21).

Considering the finite nature of freshwater resources on the one hand, and increasing demand, on the other, the need to protect and manage water resources properly is crucial. This objective of conserving and managing water properly can be accomplished only if all people of the world hold congenial perspectives and attitudes towards water resources and the governments adopt and implement appropriate policies.

Fig. 21 The great demand for water. http://1.bp.blogspot.com/-wnnmebPB5zQ/VEkxupJE5JI/AAAAAAAAEsM/ECGp38xDZQI/s1600/10629628_1488518104749018_2412422837339993091_n.jpg

Bibliography

Armstrong K (2009) The case for God. Random House, London

Cruden A (1977) Crudence complete concordance to the old and new testaments. Lutterworth Press, Cambridge, UK (first published in 1930)

Douglas JD (1962) The new Bible dictionary. Inter-Varsity Press, London

Gehman HS (1974) The new Westminster dictionary of the Bible. Westminster Press, Philadelphia (Second Printing, 1974)

Lewis N, Reinhold M (1955) Roman civilization: source book II: the empire. Harper Torch Books, New York

Nemet-Nejat KR (1998) Daily life in ancient Mesopotamia. Hendrickson Publishers, Peabody, MA

Sarkar RL (2000) The Bible, ecology and environment. ISPCK, Delhi

The Bible: Revised Standard Version

Ward B, Dubos R (1972) Only one earth. W. W. Norton & Company, New York

World Commission on Environment and Development (1987) Our common future. Oxford University Press, Delhi

Websites

http://globalministries.org/resources/special-days/IDW/

http://www.united-church.ca/ecology/water

www.un.org/waterforlifedecade

www.oikumene.org/en/activities/ewn-home.html

www.quality-drinking-water.com/bible_and_water.html

Documentary—FLOW: For the Love of Water, A Steven Starr Production

Water in Hindu Scriptures: Thank You, Water!

Sudhakara Sharma and M.S. Shruthi

Introduction

Water scarcity is affecting every continent of the globe. About 43 countries with 700 million people are witnessing water crisis, and Ethiopia takes the lead followed by Haiti and Niger where water availability is least. The UN computes that although the Millennium Development Goals for clean water supply are achieved, about 800 million people will still lack proper access to safe drinking water by 2015. And 1.8 billion people will still not have access to basic sanitation. A sustained growth of population, demand for better livelihoods and urbanisation are generating a steadily intensified water crisis worldwide. Most of the 60 million people added to the towns and cities of the world every year move to informal settlements. Women spend 200 million hours a day collecting water. The major factor determining the economic growth of several countries will be access to water resources; hence, water may assume the role of prime factor behind the wars fought in the world in future. About two million tons of waste per day is discharged to receiving waters comprising human waste, industrial wastes and chemical and agricultural wastes. More than 3.4 million people die each year globally of water, sanitation, and hygiene-related causes and degraded environment.

Growing population is causing major stress on all the natural resources. Sewage and agricultural runoff has contaminated most of the water resources. Agriculture has always been a major consumer of water and accounts for around 70% of water used. Unsustainable agricultural practices are forcing farmers into resorting to undue exploitation of water. Overuse of water in agriculture is also causing land degradation threatening future food security and livelihoods. Local varieties which are geo

S. Sharma
Independent Sanskrit scholar
e-mail: sudhakarasharma@gmail.com

M.S. Shruthi (✉)
Research Assistant, Institute for Social and Economic Change (ISEC), Bengaluru, India
e-mail: shruthi_msm@yahoo.com

K.V. Raju, S. Manasi (eds.), *Water and Scriptures*,
DOI 10.1007/978-3-319-50562-6_4

specific and season specific are ignored. Further increased urbanisation has added to the complexity in cities facing scarcity, pollution, inequity, etc. Water, sanitation and hygiene-related causes result in morbidity and mortality concerns besides degraded environment. There are several trans-boundary problems due to water scarcity, like Sindhu River dispute between India and Pakistan; dispute over the sharing of Nile waters between Egypt and Sudan; the issue among Thailand, Vietnam, Lao People's Democratic Republic (Lao PDR), Cambodia, China and Myanmar pertaining to Mekong River Basin; dispute between India and Bangladesh over the sharing of Ganges waters, Egypt and Ethiopian conflict over Nile and several others.

The purpose of this paper is to illustrate from Hinduism thoughts that can help promote conservation and proper water management. The role of culture plays a significant role in effective water resources management. Identifying values in Hindu scriptures is the purpose of the paper, besides drawing lessons to aid current challenges of water management. The methodology adopted is through explaining the religion in brief as protecting nature is part of Hinduism and explaining the various literature sources from where we have sourced the values and principles drawn regarding the precious resource water.

Hinduism

Hinduism is a path of life being practised from time traced back to 10,000 BC as per the records available to the mankind till date. Hinduism is more of a way of life than a religion as this path was not propounded by a single person or an incident, but has just been naturally practised by humans from time immemorial; thus, the religion is described as '*Sanatana Dharma*, the path which does not have a beginning'. As such Hinduism keeps itself away from fundamentalist attitudes, while embracing virtues from anywhere in the universe.[1] Hence, the conception of the whole world as 'one family'[2] is derived, making Hinduism the most pacifist faith in the world.[3]

Ecology and Hinduism

The pacifism emitted by Hinduism does not get confined just to the mankind, but extends very much to the well-being of every living and inert being in the whole universe as everything in nature is equated with the almighty.[4] *God has manifested himself through the creation of this universe,* so man is forbidden from exploiting

[1] आनोभद्राः क्रतवो यन्तु विश्वतः/Aanoibhadraah krathavo yanthu vishwathah

[2] वसुधैव कुटुम्बकम्................ Vasudhaiva kutumbakam……Hithopadeshah

[3] सर्वे भवन्तु सुखिनः, सर्वे सन्तु निरामयाः, सर्वे भद्राणि पश्यन्तु मा कश्चित् दुःखभाग्भवेत्/ Sarve bhavanthu sukhinah sarve santhu niraamayaah sarve bhadraani pashyanthu maa kashchit dukha bhaagbhavethh, sarve bhadrani pashyantu maakashchit dukhkha bhaagbhaveth

[4] सर्वम् खल्विदम् ब्रह्म/sarvam khalvidam brahma

nature and is bound to live in harmony with nature and recognise the divinity that prevails in all elements, including plants and animals.[5]

This is the philosophy behind Hindus worshipping everything. As such, every rite and ritual performed through this path chants coexistence with the nature, as nature happens to be the very base of existence.[6] Thus, Hinduism has always been an ecologically insightful idea. Environmentalism is an intrinsic part of a divine world outlook in Hinduism.

Hindu Scriptures

As delineated, Hinduism, being a way of life, has branched out itself over time into various subjects and specialisations to comprehend the almighty and help humanity live in coherence with nature in all possible ways and attain the ultimate bliss.[7] The knowledge of this apparent/mortal world is termed **Avidyaa, aparaa vidya** (अविद्या/अपराविद्या) and the knowledge of the cosmic world is termed **Vidyaa, Paraa vidyaa** (**विद्या/पराविद्या**).[8] **The paraa janana** (परा ज्ञानम्/विद्या) has been given more emphasis but with no negligence assigned to the life in the mortal world or without compromising on the worldly life.

Vedas are the foundation rocks on which the structure of Hindu culture, life and scriptures have evolved. Vedas are the most authentic and the most ancient literature available about Hinduism. Etymologically, 'Veda' comes from the root verb 'Vid' which means knowledge; Vedas are thus a compilation of knowledge meant for guiding mankind.

The Vedas are the direct revelations made to the mankind by the supreme power as there seems to be no evidence of it being written by anybody nor is it possible for a human brain to put forth knowledge in such a perfect order.

Vedas comprise hymns fragmented into four broad chapters,[9] namely, Rig, Yajus, Sama and Atharvana, based on the themes of hymns. The Rig Veda emphasises knowledge; the Yajur Veda emphasises duties; the Sama Veda emphasises devotion; and the Atharvana Veda emphasises physical sciences.

---

[5] आप्रारजांसि दिव्यानि पार्थिवा.............प्रबाहू अस्राक् सविता सवीमनि निवेशयन् प्रसवन्नक्तुभिर्जगत्...............(ऋक् ४.५३.३)/aapraa rajaamsi divyaani paarvthiva........prabaahu asraak savitaa savimani niveshayan prasavannaktubhirjagat.......Rigveda...4.53.3

[6] त्वं विभर्षि द्विपदः त्वं चतुष्पदः.....................अथर्वणवेदः १२.१.१५/tvam bibharshi dvipadah tvam chatushpadah….Atharvanaveda…12.7.15

[7] मोक्षः/Mokshah.

[8] अविद्यया मृत्युं तीर्त्वा विद्यया अमृतत्वमश्णुत.................................. Avidyayaa mrutyum teertvaa vidyayaa amrutatvamashuta....Yajurveda...40.11

[9] ऋग्वेदोऽथयजुर्वेदोसामवेदोह्यथर्वणवः/Rigvedothayajurvedahssaamavedoatharvanah

Each of the four Vedas has its own *Samhitaa (a collection) which is, as said above, the direct revelations made to the mankind by the supreme power and is the most valid part of Hindu literature.*[10] Apart from 'Samhitaas' each Veda has been developed by human efforts with branches called 'Brahmanas', 'Aaranyakaas' and 'Upanishads' which are generally the records of discussions of the 'Samhitaas'.

An Upaveda exists for every Veda. Ayurveda, the science of life, for Rig Veda; Dhanurveda, the art of disaster management, for Yajur Veda; Gandharva veda, the science of fine arts, for Saama Veda; and Arthaveda, the science of wealth, for Atharvana Veda.

There are six Vedangas or the parts of Vedas called Vyakarana, grammar of Vedas; Chandas, the prosody in Vedas; Niruktham, the thesaurus of Vedas; Shikshaa, the phonetics in Vedas; Jyoutisha, astronomy; and Kalpa, the subject dealing with the ceremonial injunctions of principles in Vedas. *Both the Upavedas and Vedangas are the products of manual work.*

Chart 1

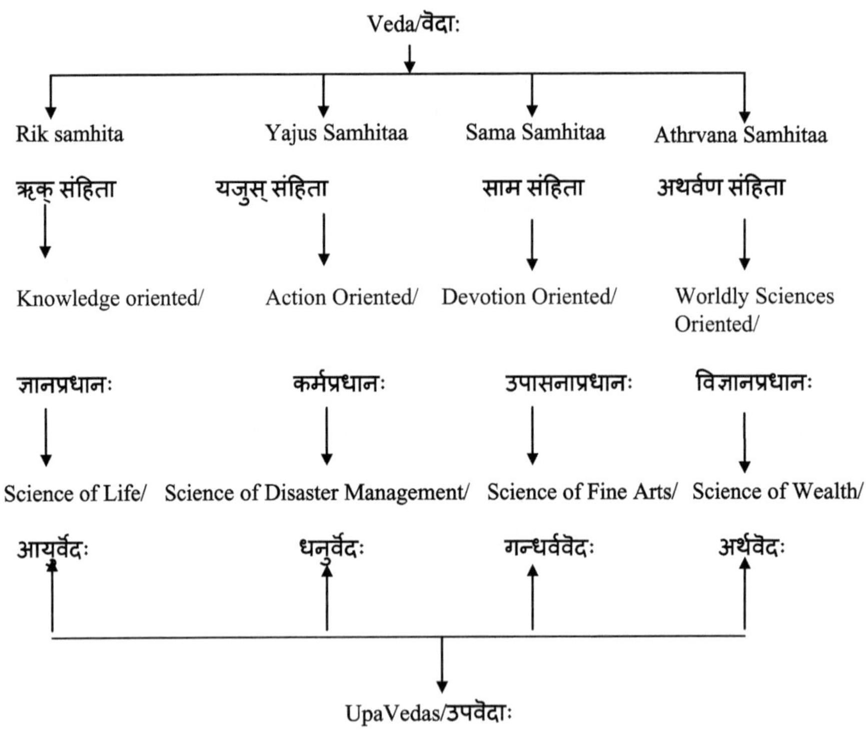

[10] श्रुतिस्मृतिविवादेषु श्रुतिरेव गरीयसि...,आम्नायप्रमाणम्, शब्दप्रमाणम्/ shruthismruthivivaadeshushruthireva gareeyasi...aamnaayapramaanam/shabdapramaanam

Chart 2 Chart of Vedangas

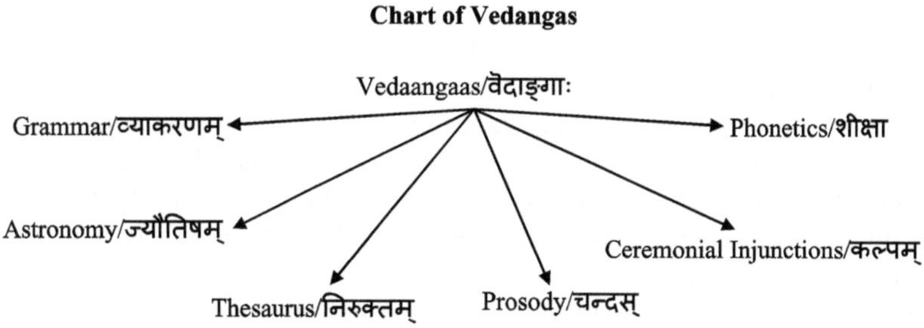

All the sciences or Vidya developed by the Hindu civilisation are the extracts of knowledge provided by Vedas. The historical records mention as many as 64 different branches of knowledge that were in practice, but today the literature of the past is available only for a few subjects. The below mentioned subjects could well fit for consideration for a study of the perspectives of water in Hindu/ancient Indian literature.[11]

Chart 3 Chart of literature considered

- Medicine[11]/Ayurveda, vaidyashastram/वैद्यशास्त्रम्
- Political Economy/Arthashastram/आर्थशास्त्रम्
- Architecture/Vaasthushastram/वास्तुशस्त्रम्
- Chemistry/Rasashastram/रसशास्त्रम्
- Law/Dharmashaastram/धर्मशास्त्रम्
- Literature/Kaavyaani/काव्यानि
- Astronomy/Jyoutisham/ज्यौतिशम्

Water in Hinduism

Excerpts from Verses: The abundant Hindu writings explain creation in various ways and contexts, but as mentioned earlier, the most authentic of the literature, the Vedas, describes 'Cosmology' through a set of hymns called Naasadiiya sooktham (नासदीय सूक्तम्....ऋग्वेदः...... 10.129) in the 10th 'Mandala' or the section of Rig

[11] Botany happens to be a part of the medical sciences, it is classically named as Vrukshaayurveda (वृक्षायुर्वेदः)

Veda. The Naasadiiya sooktham states figuratively that there was nothing called nothing/null/vacuum at the time of creation.[12] It clearly states that God did not craft this universe from nothing. There was something called *Ambah*,[13] the eternal and the most primordial state of elements.

Hindu or the Vedic philosophy suggests the presence of five constituents or elements 'Pancha Bhootas'; they are earth, water, fire, air and ether.[14] These elements emerged out of the 'Ambhah', one following the other. This proves that Hindu ideology had inferred that matter could neither be created nor be destroyed but only could transform, in the very early period of human history. Vedas describe that everything in this world is a mixture of the above said five elements in different proportions.[15]

> Modern science speaks of 108+ elements which constitute the universe and are categorised according to their properties/features in the periodic table. These have been described under five major heads in the ancient Indian sciences thousands of years well before the construction of the periodic table.

The Ambhah advanced from ether to form to air; air emitted fire; fire moved on to shape water which, in turn, formed earth; and earth gave birth to life.[16]

Fire moved ahead to form water at 'Apraketa' (अप्रकेत) the subtle and 'Prakeeta' (प्रकेत)[17] the gross state. The 'Praketa' or the gross state of water is what is available to view, and hence, water is referred to as *Pancheekrutha tatvam.* This comes close to the proposition of modern sciences of water being a combination of two atoms of hydrogen and one atom of oxygen (H_2O).

The origin of the universe has been traced to water itself in many of our ancient texts. In the Taittiriya **Samhita**,[18] the existence of water in the beginning of creation is acknowledged, and similar declarations are made also in Satapatha Brahmana and Taittiriya Brahmana.

---

[12] नासदासीत् नोसदासीत्....................................(ऋग्वेद: 10.129.1)/naasadaasiit nosadasiit.... Rigveda...10.129.1

[13] अम्भः/Ambhah

[14] इमानि पञ्च भूतानि पृथिवी वायुः आकाशः आपज्योतीषि............ऐतरेयोपनिषत् ३-३/Imaani panchabhootani pruthivi vaayuh aakaashah aapajyotishi…..Aitareyopanishat….3.3 ???

[15] sarvadravyam paanchabhoutikamasminnarthe….Charakasamhitaa

[16] पृथिव्योष्धयः...... / tasmaadvaa etasmaadaatmana aakashassambhootah aakaashadvaayuh vaayoragnih agneraapah adbhyah pruthivi prutivyoshadayah…..taittariyopanishad….2.1.1

[17] तम आसित्तमसा गूढमग्रे प्रकेतम्। सलिलं सर्वदा इदम्। तुच्छेनाभ्वपिहितं यदासीत्। तमसस्तन्महिना जायतैकम॥/ tama aasittamasaa gudhamagre praketam. Salilam sarvadaa idam. Tuchchenaabhvapihitam yadaaset. Tamasastanmahinaa jaayataikam……Rigveda

[18] Taittariya samhitaa and vaajasaneeyi samhitaa are two samhitaa texts available for Yajurveda, of which taittariya samhitaa is not checked with prosody or chandas, a vedaanga and the vaajasaneyi samhitaa is bound by meters so could be concluded as more authentic. Taittariya samhitaa is popular in South India and Vaajasaneyi samhitaa in North India.

Several Upanishads[19] advocate 'The core of all creatures is the element earth and the essence of earth is water'.[20] 'The element earth sustains all creatures and the earth is sustained by water, the water gets transformed into herbs and vegetations, they in turn become flowers and then fruits and fruits support the creatures'[21] 'the water is the main ingredient of herbs and plants (in particular human beings)…'.[22] Every food has water as its constituent, and many times water will be a major constituent; hence, 'water is food'.[23] Brhadarhyaka Upanishad and 'Jaimini Upanishad' also say water was the primary element of creation.[24]

In 'Manusmriti', the book of law, Manu also cites 'water created the life'.[25] This is supplemented by Bhagavad Gita which says all living beings originate from food; the food is a creation of rain that is water; rain is a derivative of Yajna and Yajna activated by Karma.[26]

In the Shwetashvataropanishat, "God is proclaimed to be the source of five elements, the building resources of the cosmos; metaphorically, the almighty himself is professed as the five elements".[27]

Hence, Vedas describes water as 'Devatha', a synonym for God or 'Paramatma', the supreme soul.[28] Water is the light, the essence, the nectar and the god, the Brahman.[29]

Perceived Benefits of Water

Jeevanam is a word given to water in Vedic language and Sanskrit, the language of Indian scriptures, spoken language of ancient India and the language that linked the whole of India till the spread of English education in the country and is still studied in every part of India.

---

[19] There are more than 100 upanishads identified, the most famous among them being Esha (ईश), Kena (केन), Katha (कठ), Munda (मुण्ड), Mandukya (माण्डुकय), Pashna (प्रशन), Taittariya (तैत्तरीय), Eteraya (एतेरय), Chandogya (छान्दोग्य) and Bruhadaranyaka(बृहदारण्यक).

[20] ऐषां भूतानाम् प्रुथिवी रसः प्रुथिव्या आपो रसः............छान्दोग्योपनिशत्...१.१.२ Aeshaam bhootanaam pruthivi rasa, pruthivyaa aapo rasah…Chandogyopanishad 1.1.2

[21] एषां वै भूतानां पृथिवी रसः पृथिव्या आपः आपां ऒषधयः ऒषधीनां पुष्पाणि, पुष्पानां फलानि,फलानां पुरुषः.......बृहदारण्यकोपनिषत्.......६.४.१/ Esham vai bhootaanaam pruthivi rasah, pruthivyaa aapah, aapaam oshadhayah oshdhinaam pushpaani pashpaanaam phalani phalaanaam purushah… Bruhadaranyakopanishat..6.4.1

[22] आप ऒषधयो वनस्पतयः......तैत्तरीयोपनिष्त्...१.७.१/ aapa oshadhayo vanaspatayah….. Taittariyopanishat….1.7.1

[23] आपो वा अन्नम्....तैत्तरीयोपनिषत्..३.८.१/ aapo vaa annam… Taittariyopanishat….3.8.1

[24] आपोवा अग्रे इदमासन्......................... aapo vaa agre idamaasan….Jaiminyopanishat

[25] आपो ऐव ससृज आदम्......................... aapo eva sasruja aadam….Manusmrithi

[26] annaabhavati bhootaani parjanyaadannasambhavah yajnaadbhavatiparjanyah yajnah karma samudbhavah…Bhagavadgeeta

[27] श्वेताश्वतरोपनिषत्..........४.२/shetaashvataropanishat….4.2

[28] आपो वै देवता............................ aapo vai devataa…Yajurveda..14.20

[29] आपो ज्योती रसोऽमृतम् ब्रह्म भूर्भुवस्सुवरोम्......................... aapo jyothi raso amrutham brahma bhoorbhuvassuvarom….Yajurveda…15.20

The word 'Jeevanam' is derived from the root verb 'jeeva' meaning 'Praanadharane', embracing life. Water is given the word Jeevanam to show its wonderful importance in life. Water is thus comprehended as the elixir of life.

Water is broadly found in scriptures as an utter necessity in 'snaana' (bathing), 'paana' (consumption), 'shoucha' (cleansing), 'chikitsaa' (relieving), 'upachaara' (hospitality), 'krushih' (farming) and 'tarpanam' (consoling).[30,31]

India, being a peninsula, trade associated with the means of water could never be ignored not just in the past but also at present. The word navigation is derived etymologically from the Sanskrit word 'navah' meaning boat. The traces of maritime trade in India date back to the Vedic age where we find the description reading 'although the water voyages are extremely dangerous the heroic people will win it'.[32] Indus valley civilisation was an advanced urban civilisation that began to show visible growth in both the length and the frequency of maritime voyages by 3000 BCE.

Pic 1 Indian yacht as revealed in the Fra Mauro map (1460). Source: http://en.wikipedia.org/wiki/File:WorldShips1460.jpg

The epic *Ramayana* depicts the presence of widespread waterways across the kingdom of Kosala; this is proven by the narration in the epic that Bharatha was supplied 500 different varieties of beautifully decorated strong boats by Guha to cross across Ganga with loads of commodities.[33,34]

---

30 स्नानपानशौचचिकित्साकृषिस्तर्पणम्.../ snaanapaanashouchachikitsaa….Ayurveda

31 snaanapaanaavagaaheshu hitamambu yathaamrutham

32 समुद्रं न सन्चरणे सनिष्यवः….ऋग्वेदः/ samudran na sancharena sanishyavah…..Rigveda

33 ते तथोक्ताःसमुत्थाय त्वरिता राजशासनात्। पञ्च नावां शतान्याशु समानित्युः समन्ततः॥ २.८९.१० Te tathoktaah samuththaaya tvarithaa raajashaasanaath| pancha naavam shataanyaashu samaanityuhu samantatah.........2.89.90

अन्याः स्वस्तकिवज्ञिेया महाघण्टधारा वराः। शोभमानाः पातकाभर्युक्तवाताः सुसंहताः॥ २.८९.११ anyaah svastikavijneyaa mahaaghantadhaaraa vaaraah| shobhamaanaah paatakaabhiryuktavaataah susamhataah.....2.89.11

34 ततः स्वस्तिकविज्ञेयां पाण्डुकम्बलसंवृत्तामाम्। सनन्दिघोशां कल्याणीं गुहे नावमुपाहरत्॥ २.८९.१२ Tatah svastikavijneyaam paandukambalasamvruttaamaam| sanandighoshaam kalyaaneem gruhe naavamupaaharat….2.89.12

Pic 2 Satellite picture of the submerged Rama Setu Bridge constructed across Palk Straight connecting India and Sri Lanka as recorded by NASA. Source: https://www.google.co.in/search?q=Images+of+rama+sethu+along+in+the+Indian+ocean&tbm=isch&tbo=u&source=univ&sa=X&ei=UgCnUsk_yZKuB_ydgMgJ&ved=0CDMQsAQ&biw=1708&bih=739&dpr=0.8#facrc=_&imgdii=_&imgrc=lDV8aNtVlC2mJM%3A%3BelZYm8FRvWRf-M%3Bhttp%253A%252F%252Fwww.hindujagruti.org%252Fnews%252Fout%252Fimages%252F1361610027_ram-setu.jpg%3Bhttp%253A%252F%252Fwww.hindujagruti.org%252Fnews%252F15743.html%3B350%3B225

The maritime trade had also developed well under the rule of the predecessors and successors of Rama. Also it is well known that Rama got a viaduct erected between the mainland of India and Sri Lanka to make it to the city of Lanka to fight Ravana for bringing Sita back.

Water in India has also been employed for composition of music through a mechanism of a set of ceramic or metal cups of various volumes. This is named *Jalatarang*. The earliest references to this apparatus can be traced from the same epic of 'Ramayanam'.

Pic 3 Jalatarang maestro in action. Source: https://www.google.co.in/search?q=pictures+of+jaltarang&biw=1600&bih=828&source=lnms&tbm=isch&sa=X&ved=0ahUKEwi3jJKF5JLNAhVJKo8KHS0EBkcQ_AUIBigB#imgrc=3KevfGqQON0glM%3A

A type of water clock called ghati[35] yantra was very much in use; traditionally this clock was used to set time for auspicious ceremonies; the bowl, with a small hole at its bottom, sinks after 24 min.

Pic 4 Ghati yantra. Source: Science in Ancient India, a Presentation by Michel Danino

Figuratively, the nature of water is used in several places. 'Water that rains into a fortress runs among the hills so does the Dharma in the form of its result/effect'.[36]

Chanakya in his treatise on conduct 'Chanakya Niti' has described the importance and benefits of water, so 'in the whole world, there are only three precious pearls: water, food and the subhshitas, the adages found in Sanskrit language'.[37]

Medicinal Properties of Water

Water, one of the five basic elements, is considered as a building material, a sustaining force and also as a curative medicine in ancient Indian culture. The earliest recorded Vedic literature, Rig Veda, describes water as the universal medicine.[38]

[35] A unit of time called ghati equal to 1/60th of a day.

[36] यथौदकं दुर्गे वृष्टं पर्वतेषु विधवति। ऐवं धर्मान्पृथक्पश्यान् तानेवानुविधावति॥..कठोपनिषत्..२.१.१४......... yathoodakam durge vrushtam parvateshu vidhavathi| evam dharmaanpruthakpashyaan taanevaanu-vidhaavathi……..Kathopanishat……2.1.14

[37] Pruthivyaam treeni rathnaani jalamannam subhashitham…..chanakyaniti

[38] Aapa idvaa u bhishjiraapo amiivachaatanih aapah sarvasya bheshajiistatre krunvanthu bheshajam….Rig 8.7.24

Ayurveda, the science of life and a derivative of Rig Veda, has embraced the Vedic principle of the universal usage, applicability and employment of water to the fullest extent. Ayurveda advocates various treatments employing water and various methods of consuming water, defines time to consume water, says how to consume and describes the various dimensions of water such as cold water, hot water, heated and cooled water, rainwater, lake water, well water, ice and so on as a medicine or otherwise.

Ayurveda describes water as the life of all living beings. The world is predominantly watery when in health or in ill health, and there can be no life without water.[39] Water is used both as an independent medicine and a base medicine. Ninety percent of the Ayurvedic medicines are concoctions of water. Water, as a medicine, has both internal and external uses as per science. Hydrotherapy, a concept of naturopathy, has got its roots in Ayurveda.

Aqua is a major chemical required for digestion of food taken in. It is advisable to sip little water during meals.[40] It is an adage prevalent in Ayurveda that during indigestion the right and suitable food is water,[41] preferably hot water. Water taken before meals will dampen the 'Jatharaagni' (the digestive power) and dilute the digestive juices, and in the long run, it results in malassimilation (ineffective assimilation). Water, when taken immediately after meal, causes obesity, and hence, it is advisable to take little water in the course of meals.[42] The same theory is propounded in other text quoting water consumed in the middle, at the end and in the beginning results in a balanced structure, obese structure and a lean structure, respectively.[43] The food doesn't get digested and assimilated if water is consumed in very high quantities. The same problem occurs when water is consumed in too low quantities. For good appetite it is essential that more water is consumed intermittently.[44]

---

[39] पानीयम् प्रणिनाम् प्राणः विश्वमेवच तन्मयं नहि तौयाद्विना वृत्तिः स्वस्थ व्याधितस्यवा............ अष्टाङ्गसङ्ग्रहः paaneeyam praaninaam praanah vishvamevacha tanmayam nahi toyaadvinaa vrittih svasthavyaadhitasyavaa.........Ashtaangasamgrahah

[40] अशीतश्चौदकं युक्ता भुन्जानश्चान्तरां पिबेत्..........................सुश्रुतसंहिता asheetashchoodakam yuktaa bhunjaanashchaantaram pibeth..........sushruthasamhitaa

[41] Ajeerne bheshajam vaari jeerne vaari balapradam bhojane chaamrutham vaari bhojanaanthe vishapradam.......Chanakyaniithi

[42] Bhukthasyaadau jalam piitham kaarshma mandaagni doshakruth madhyegni deepanam shreshthamanthe sthoulya kaphapradam

[43] Samasthoolakrushaabhuktha madhyaantha prathamaambupaah.......Ashtaangahrudayam Sutrasthana 5

[44] अत्यम्बुपानान्न विपच्यतेऽन्नं निरम्बुपानाच्च सयेव दोषः तस्मान्नरो वन्हिविवर्धनाय मुहुर्मुहुर्वारि पिबेद्भूरि....……...सुश्रुतसंहिता atyambupaanaanna vipachyatennam nirambupaanaachcha sayeva doshah tasmaannaro vanhivivardhanaaya muhurmuhurvaari pibedvaari........sushruthasamhitaa

Water taken at dawn works like the heavenly nectar, and in the process of assimilation, it bestows strength; water works like poison when taken immediately after food and as a medicine when properly employed during disease condition.[45]

Drink water at the end of night/daybreak[46] says Ayurveda; this is widely referred to as 'Ushah Paana' (उषः पानम्) (drinking of water during dawn), performed as a routine cultural practice by Hindus.

Astaangahrudayam, a prominent ancient text of preventive medication written by Acharya Vagbhata, prescribes water to be sipped not drunk at once; water should be consumed at least 45 min before food and 90 min after food.

Charaka Samhita written by Charaka also substantiates the statements made in Ashtangahrudayam, while stating that water becomes toxic when consumed at the end of a meal.

Sushruta Samhita written by Sushruta also says that water becomes toxic when consumed in too much quantity, and it is also toxic not to have water at all.

Ayurveda identifies three main defects in the human body[47] called 'vata dosha', 'pitta dosha' and 'kapha dosha'; a balance of vata, pitta and kapha indicates a normal/good health of the body; and an imbalance of them indicates ill health termed with the said doshas or defects. For those who suffer from 'vata dosha', Ayurveda advises swigging hot water, and for those anguishing from 'pitta dosha', drinking cool water is recommended.[48]

45 उषःकाले अमृतम्वारि जीर्णेवारिबलप्रदम् विषवद्भोजनान्तेच भैषज्यम्रोगपीडिते...............चरखसंहिता Ushah kale amrutham vaari jeerne vaari balapradam vishavadbhojanaanthe cha bhaishajyamroga-pidithe………Charakhasamhitaa

46 निशान्तेच पिबेद्वारि……………Nishaantecha pibedvaari

47 वातपित्तकफदोषानि त्रीणि/ Vaathapiththa kaphadoshaani thriini

48 Ushnam vaathe kaphe piththe raktha cha shiithalam ………Sushrutha Sutraani….46.435

Warm water restrains the imbalances of vatu and kapha, releases fat, increases the volume of the urine excreted (vasti shodhak) and dispels or reduces fever. It is favourable in the cases of cough and difficulties of breathing and is nourishing always.

Water boiled and reduced to a quarter of the initial quantity and then cooled removing its froth and ebullition would be light, clear and safe for everyone and helps balance the tri defects (vata, pitta and kapha) of the body. It is also known to be beneficial in treating diseases acquired by overconsumption of wine, dysentery, skin burns, bloody mucus, effects of any absorbed toxins, vomiting, catarrh, vertigo and unconsciousness.

Water boiled in the previous night is not recommended for quenching thirst as it would probably have acquired acidic taste and may cause kapha defect. The tender coconut water is grave and soothing, cool, enjoyable and delicious. It enhances urination, secretes semen and desists thirst and pitta defect.

Cold water is known to be helpful in treating epilepsy, in summer, in the condition of excessive body heat, the imbalance of Pittam, treating blood poisoning, problem associated with excessive consumption of wine, the state of unconsciousness, exhaustion, vertigo or dizziness, tamaka and nausea. Although cold water is good and is recommended to be used as medicine, its use is not advised under conditions, such as pain at the sides (pleurodynia), catarrh, rheumatism, diseases of the larynx, distention of the stomach by gas or air, cases of undigested faeces, acute stage of fever, just after the exhibition of any emetic or purgative remedy, severe cough and soon after consuming fatty or oily drinks or Snehapaana acute cold, vaatha diseases, sore throat, gastritis, constipation, fever immediately after dysentery and nausea, during hiccups and on consuming more of oily food.[49]

Minimum consumption of water is advised for a person suffering from loss of savour for food, heartburn, oedema, any of the wasting diseases, weakened digestion, abdominal dropsy, skin diseases, fever, diseases affecting the eyes, ulcer and diabetes (Madhumeha, etc).[50]

As already stated, in addition to internal use, Ayurveda also recommends the external use of water as a way of maintaining good health and also as a therapy. Ayurveda mandates that every human to bathe every day.[51] This is also a natural requirement in India as the weather conditions are mostly hot and humid throughout

---

[49] पार्श्वशूले प्तश्याये वातरोगे गलग्रहे। अधमाने स्थिमिते कोष्टे सद्यः शुद्धे नवज्वरे। हिककायां स्नेहपीतेच शीतांबु परिवर्जयेत्॥ सुश्रुत सम्हिता सूत्रश्तन ४५....Parshwashoole prathishyaaye vaatharoge galagrahe। Aadhmaane sthimithe koshte sadyah shudde navajware hikakaayaam snehapeethe cha sheethaambu parivarjayeth (Sushruthasamhitha. Soothrasthana. 45.)

[50] An extract from the Sushruta Samhita edited and published by Kaviraj kunja lal bhishagratna with a full and comprehensive introduction and plates (in three volumes) Kalyanaraman.

[51] दीपनम् वृष्यमायुश्यम् स्नानमूर्जा बलप्रदम् कदूमलाश्रमस्वेदतन्द्रा तृत् दाह पापजित्/Deepanam vrushyamaayushyam snaanamurjaa balapradam kadumalaashramasvedatandraa truth daaha paapajith

the year. Further, it recommends massaging of the body and head with oil before bath.[52] A bath increases appetite, gives strength, adds years to life, enhances agility, nourishes and helps overcome itching, toxins, fatigue, sweating, lethargy, thirst and heat.

'Foot bath' is generally understood as a curative measure in naturopathy, whose origin is attributed to other than Indian. But precise reference is there in Ayurveda texts about bath. Foot bath improves eyesight and pacifies the mind.[53] This rejuvenates the circulatory system. This foot bath is recommended for curing acute headache, insomnia, disorders related to blood pressure, etc.; for a hot hip bath, the water should be around 40°C and the duration is 10 min.

Water is just H_2O at all places and at all times. But Ayurveda takes a different view. The quality and properties of water change in a subtle way depending on the season.

The seasonal impact on water is propounded as follows. 'Although water in the spring (वसन्त ऋतु) season is sweet, owing to the generation of an astringent taste,[54] it has a sort of dry quality'.[55] In the summer (ग्रीष्म ऋतु) season, as temperature swells, water shuns surplus emanations in the body.[56] 'The water in the rainy (वर्ष ऋतु) season is harder and causes secretions, in addition to being sweet and fresh'.[57] The same water during autumn (शरत् ऋतु) season satiates better, is light and doesn't aggravate secretions in the body.[58] In the snowing (हेमन्त ऋतु) season, water is relatively heavy and bestows lubrication, power and strength and enhances comfort.[59] In the winter (शिशिर ऋतु) season, heat in the water is very little, and it will be light and conquer kapha and vaatha.[60]

---

[52] अभ्यङ्ग स्नानम्/Abhyanga snaanam

[53] पादावगाहनेन नेत्रपाटवः मनःशान्तिश्च/ Paadaavagaahanena netrapaatavah manah shaanthishcha

[54] **ಒಗಚು/Ogachu**

[55] कषाय मधुरं रूक्षं वासन्तिकं जलम्.....चरक संहिता सूत्रस्थानं २७, Kashaya madhuram rooksham vaasanthikam jalam. (Charakasamhitha.Soothrasthana.27)

[56] गरिष्मिकं तु अनभिष्यान्दि जलमित्येव निश्चयः.... चरक संहिता सूत्रस्थानं २७ Graishmikam thu anabhishyandi jalamithyeva Nishchayaha (Charakasamhitha.Soothrasthana.27)

[57] गुर्वभिष्यन्दि पानीयं वार्षिकं नवं....... चरक संहिता सूत्रस्थानं २७ Gurvabhishyandi paaneeyam vaarshikam madhuram navan. (Charakasamhitha.Soothrasthana.27)

[58] तनुलाघवन्भिष्यन्दि प्रायः शरदि वर्शति..... चरक संहिता सूत्रस्थानं २७ Thanu Laghvanbhishyandi praayaha sharadi varshathi (Charakasamhitha.Soothrasthana.27)

[59] हेमन्ते सलिलं स्निग्धं वृश्यं बलहितं गुरु..... चरक संहिता सूत्रस्थानं २७ Hemanthe salilam snigdham vrushyam bala hitham guru (Charakasamhitha.Soothrasthana.27.)

[60] किञ्चित्तप्तो शिशिरे कफवातजित्...... चरक संहिता सूत्रस्थानं २७ Kinchith taptho Shishire kaphavaathajith (Charakasamhitha.Soothrasthana. 27.)

Fasting once in a fortnight, preferably on the 11th day, Ekaadashi, is a part of religious practice. No food is recommended for consumption on that day. Fasting to cleanse the body happens to be the spirit of the practice. In either case fasting turns out to be beneficial when sufficient water is consumed. The water consumed cleanses the entire gastrointestinal trail and eliminates the lingering faecal stuff from the large intestine and the rectum thus checking putrification. This improves appetite and reduces the toxic condition in the body. It also helps one to control one's tongue and gluttony (palate).

The attributes of rainwater gathered prior to the contact with land are listed by Sushruta in the 45th branch of suthrasthana in his samhitha. 'It beats the disparities caused by Vaatha, Piththa, Kapha; offers vigor, augments the seven building materials of the body known as sapthadhaathus, enhances the brain activity. Once it touches the land its quality changes according to the quality of the terrain'.[61]

The water of rivers, (which drain the Jangala, desert countries) flowing into the western sea is light and, therefore, wholesome. The river waters that cross marshy (Anupa) areas and join the eastern oceans are heavy, and as such its use is restricted to minimum since they increase the oily secretions of all the organs. The water of rivers, which runs into the southern sea, is neither too heavy nor too light owing to the fact of its traversing countries which have a 'Sadharana' (ordinary) character.

The waters sourcing from the Sahyadri mountains are unsuitable for the skin, while the water originating in the mounts of Vindhyas causes leprosy and Jaundice. The water of rivers, which rises in the Mount Malaya, begets worms and intestinal vermin, while the water sourced from the Mahendra Mount begets elephantiasis and abdominal dropsy. The water of rivers, which rises in the Himalaya, produces angina pectoris, Hridroga, anasarca, diseases of the head, elephantiasis and goitre. Similarly, the water, which drains the eastern part of the land of Avanti or flows across its western fraction, begets piles, while the water that rises in Paripatra Mount is nourishing, strength giving and favourable to health. The water in a Jangala country is free of baneful traits, ideal and acidic in digestion (Vidahi), possesses commendable qualities and is pleasing and cooling. The water of a Sadharana, a temperate country is light, cool, enjoyable and tasty or Dipanam.

The waters of swift-moving and clear rivers are light and that of slow-moving and enclosed with mosses and marine growth and are weighty. The river waters of the Murudesha, presently Marwar in Rajasthan, will be saline, bitter or sweet tasting and are easily digestible and strength giving.

[61] गगनाम्बु त्रिदोषघ्नं गृहीतं यत् सुभाजने। बल्यं रसायनं मेध्यं पातापेक्षी ततः परम्॥ Gaganambu thridoshagnam Griheetham yath subhaajane।Balyam Rasaayanam medhyam paathaapekshee tathah param॥

The lake water quenches thirst, gives strength, tastes sweet and is also harsh. The pond or tank water produces Vayu, tastes sweet and is harsh and strong for digestion. The water from a large tank or Vapi supresses the imbalances of Vayu and Kapha and causes Pitta defect, is strong in taste and will be charged with alkaline solution. The waters of Chunti are good for digestion, cause dryness and taste sweet. The well (Kupa) water is appetising and causes Pitta defect. The fountain water will be light, appetising and enjoyable to drink and destroys Kapha defect in the body. The Artesian spring water tastes sweet and restricts Pitta defect. It acts as an antacid for digestion. The water of Vikira is supposed to be light, strong in taste and alluring and charges with Khara or potash.

Water which gets collected in an open area or on fallow area is grave for digestion and augments the tri defects of the body. So does the water of Palvalam, a king of waterbody. Seawater smells fishy, tastes saline and aggravates all the defects in the body.

All kinds of surface water must be collected early at dawn as that part of the day is coolest and fresh; these two attributes by far from the most commendable traits in water.

Waters getting in contact with the sunlight in the day and reflecting moonlight in the night doesn't cause Kaphdosha or dryness, and as such it is equivalent to the purest form of rainwater. Rainwater collected in good container holds the virtue of suppressing the three imbalances of vaata, pittah and kapha in the body and behaves as an elixir.

The cool and limpid washings of the gem known as the Chandra kanta Mani (the moonstone) is believed to possess the supernatural powers of warding off the strikes of beasts and evils and subdues the imbalance of Pittam. They are beneficial in fever and poisoning cases marked by a burning sensation of the body, etc.

Prayers on Water

Worshipping is the most significant part of Hindu faith. The Hindu school of thought emphasises that submitting oneself and expressing gratitude to the almighty is the most vital of routines of every human being. The directives revealed in Veda Samhitaas or any other text is mostly in the form of prayers. Hence, 'Prarthana' or praying is a canon for a Hindu, and thus, 'Prarthana' is interwoven with life completely by Hindus who pray to everything everywhere and at every step of life.

But prayer, as is presumed in general, is not a moment chosen to beg God to bless us with what we need, but actually a prayer is a declaration to commit oneself to do what is required to achieve a need.[62] Hence, everyone is required to take a pledge do the best to the nature everyday while praying for receiving all that we need from it.

---

[62] प्रार्थना वै संकल्पः..........................निरुक्तम्/praarthanaa vai sankalpah.............Niruktham

Water, a very beautiful and precious part of nature, is revered with utmost devotion. There are innumerable hymns, statements and sayings in Veda Samhitaas and works affiliated to them such as Brahmanas, Aranyakas and other mythical works, such as Puranaas devoted to value the blessings that water bestows on living and command to preserve and safeguard it.

A very commonly used hymn in the Yajurveda states, 'Oh water you are a confirmed giver of happiness. For food and strength and to equip us to face the realities of life and to make us deserve happiness, please nourish us. Oh water, you fill the needed life into the food we eat, herbs we use. You are indeed our life giver through the food we partake. Let our creativity be enriched consuming you. The loving mother nourishes her progeny with food, milk and what not, so you are, oh water we shall become strong consuming you. Oh water, you want your abuser/enemy to get defeated I shall defeat such efforts wholeheartedly. Please bestow on me the best of powers'.[63]

Further, Yajurveda quotes, 'Let the water on earth, in the herbs, water in the clouds, the lord of water the paramatma guide me,[64] let the thundering clouds get converted into the sought after rain'.[65] It extends further and says, 'let the water be divine and peaceful which can fulfill our wishes, nourish us and be available for our consumption, let it shower on us from all sides to provide us health and happiness.[66] Let the water be helpful to us'.[67]

Rig Veda states, 'let the oceans be calm and peaceful and let the flowing rivers and water bodies bestow on us peace.[68] Let the clouds that move in the skies be peaceful.[69] Let the rains be helpful to us, the mankind'.[70]

---

[63] ओम् आपोहिश्ठामयोभुवस्तानऊर्जेदधातन महेरणायचक्षसे योवःशिवतमोरसःतस्यभाजयतेहनः उषतीरिव मातरः तस्माऽरङ्गमामवो यस्यक्षयाय जिन्वथ आपो जनयथाचनः(य़जुर्वेदः- ३६.१४,१५,१६)/ Om aapohishtaa mayo bhuvastaana oorje dadhaatana maheranaaya chakshase | yovah shivatamo rasastasya bhaajayatehanah ushateeriva maatrah tasmaa arangamaamavo yasyakshayaaya jinvatha | Aapo janayathaa cha nah |.............Yajurveda.....36.14,15,16

[64] payah prithivyaa paya oshadhishu payo divyaantharikshe payothaah payaspathii pradishaasanthumahyam…...Yajurveda..18.22

[65] शं नः कनिक्रदद्देवः पर्जन्यः अभिवर्शन्तु (यजुर्वेद ३६.१०)/ sham nah kanikradaddevah parjanyah abhvarshantu….Yajurveda..36.10

[66] शन्नोदेवीरभिष्ट्य आपो भवन्तु पीतये, शय्योरभिस्रवन्तुनः (य़जुर्वेद ३६.१२)/sham no devirabhishtaya aapo bhavantu pitayeh shayyorabhisravanthunah……..Yajurveda…36.12

[67] आपः शन्तिः(यजुर्वेद ३६.१७)/Aapah shanthih…..Yajurveda 36.17

[68] शं नः सिन्धवः शमु शन्तौ आपः (ऋक् ७.३५.८)/sham nah sindhavah shamu shantau aapah………Rigveda…7.36.8

[69] श् नः अहिर्बुध्नियः (ऋक् ७.३५.१३)/sham nah ahirbhudniyah……Rigveda…8.35.13

[70] शं न: पर्जन्यः भवन्तु प्रजाभ्यः (ऋक् ७.३५.१०)/sham nah parjanyah bhavanthu prajaabhyah….Rigveda…8.35.10

'The earth which is filled with deep-seas, brooks and straits whose cultivators yield victuals and survive in concert; where each alive being toils with vitality; such our motherland may rear us with her juices' states the Atharvanaveda.[71]

Shatapatha brahmana terms water as nectar.[72]

We find plenty of Vedic literature such as Parjanya Sooktham,[73] Varuna Sooktham, Prithivi Sooktham, Bhoo Sooktham and Aghamarshana Sooktham dedicated to studying and praying to water. Another such prayer, also referred to as Vedic National anthem, calls for timely rain and the earth to be full of vegetation.[74]

A rite called 'Udakashanthi' is observed along the length and breadth of India which is meant for creating a peaceful atmosphere, praying for timely and adequate rains.

Parjanya Yaaga is another religious ceremony conducted to invite rains through yagnas and yaagas that help in seeding of clouds.

Coconut plays a very important role in the Hindu worshipping. Coconut contains the purest form of water[75] that will be used as 'Kalasham' in which coconut will be placed on a pot containing water[76] adorned with beetle leaves or mango leaves; this arrangement occupies a very special place especially during festivities.

The initial habitat of Indians has been the basins of the seven rivers or the Saptha Sindhavah, and these rivers are Sindhu; the five rivers of Punjab, namely, Vitasta (Jhelum), Asikni (Chenab), Parushni (Ravi), Vipasha (Beas) and Satudru (Sutlej); and Saraswathi (Muller 1890, 12); the river Saraswathi is believed to have been buried underground due to certain geomorphological changes in the region.

---

[71] यस्यां समुद्र उत सिन्धुरापो यस्यामन्न कृष्तयः संबभुवुः यस्यामिदं जीवन्ति प्राणदेजत् सा नो भूमिः पूर्वपेये दधातु.............अथर्व वेद कांड १२/yaasyaam samudra utha sindhuraapo yasaamanna krushthayah sambabhuvuh yaasyaamidam jeevanthi praanadejat saa no bhoomih poorvapeye dadhaathu....Atharvaveda..Kanda 12

[72] अमृता वा आपः.............शतपत ब्राह्मण १-९-३-७/Amruthaa va aapah........Shatapathah Brahmana...1.7.3-7

[73] Parjanyaayapragaayatha………….Rigveda…..5.6.2

[74] काले वर्षतु पर्जन्यः पृथिवी सस्यशालिनी /Kaale varshathu parjanah pruthivi sasyashaalini

[75] Naarikelodakam snigdham svaadu vrishyam himam laghu trushnaapiththaanilaharam deepanam basthishodhanam…….Ashtaangahridayam…5.19

[76] कलशम्, कलशपूजा/kalasham/kalashapoojaa

The satellite imagery indicates that the Sutlej, a tributary of Sindhu, was feeding saraswathi, which flowed through Rajasthan. On the banks of these and many other rivers of India, many hymns are chanted to show gratitude towards the waters, to conserve them and to pray for their peaceful flow.

Pic 5 Person engrossed in the ceremonial drills on the bank of a river. Source: https://www.google.co.in/search?q=Priest+engrossed+in+the+ceremonial+drills+on+the+bank+of+a+river&biw=1024&bih=622&source=lnms&tbm=isch&sa=X&ved=0ahUKEwjswP739arNAhVFGpQKHdVBDnsQ_AUIBigB&dpr=1#tbm=isch&q=a+hindu+priest+performing+the+sandhyavandana+on+the+bank+of+a+river&imgdii=gNRVdfYVXsJ8dM%3A%3BgNRVdfYVXsJ8dM%3A%3BBVaUReN6SgY5bM%3A&imgrc=gNRVdfYVXsJ8dM%3A

Water is remembered also in the literary works in fine descriptions, and it also happens to be a mandate to make a mention of water in any literary work. Kalidasa, in his drama 'Abhijnanashakuntala', says water is the first of all creation.[77] Maagha, in his Shishupaalavadham, says water was the first to be given birth.[78] Bana, in his 'kaadambari', recognises, 'it is trough water one can attain sanctity'.[79] Also there are a number of works available glorifying water, the most famous among them being Gangaastava of Shankaracharya and Ganglahari of Jagnnatha Pandita.

In the tradition of music, several tunes or ragas are devoted to invoking rains, the most famous being Amruthavarshini, Devaamruthavarshini in Carnitic stream and Meghamallala in Hindustani stream.

It should be specially noted that to achieve the happenings of all the above, a human being stands responsible to do all that those events require to happen. It would simply be a waste to pray for rains while razing down all the forests and

[77] यासृष्टिस्सृष्टिराद्याः.............................. अभिज्ञानशाकुन्तला/ yaasrushtissrushtiraadyaah...... Abhijnyaanashakuntalaa

[78] पूर्वमेव किल सृष्टवानापः तासु वीर्यमनिवार्यमाददौ..........................शिशुपालवधम्(१४.६७)/poorvameva kila srushtavaanaapah taasu veryamanivaaryamaadadau......Shishupaalavadham..14.68

[79] AaÉxirÉÉãSrÉå ÌuÉzÉÑÎ®È.........MüÉSqoÉUÏ / Agastyoodaye vishudhdhihi......kaadambari

cutting down trees. It would be totally irrational to beg God to shower prosperity while polluting the water bodies. Often atheists blame or express their mistrust in God for not delivering justice and being a mute spectator of the chaos caused in the world, but the critics conveniently forget the misusing of nature man often resorts to due to his greed and ignorance. It is necessary to understand nature as a cohesive though complex system, and a deliberate damage caused to it will result in calamities, as a consequence. It is as simple a rule as to keep oneself healthy; good practices and following a good lifestyle are a need, but if not done, one succumbs to diseases.

Qualities of Water

In the Vedic science, the whole universe is understood to be made up of the five basic elements, namely, space,[80] air,[81] fire,[82] water[83] and earth.[84] The characteristics of these five elements have been defined by the ancient philosophers as sound,[85] touch,[86] visibility,[87] taste[88] and smell,[89] respectively. These elements are inert in nature and are perceivable by our sense organs, i.e. the ears, skin, eyes, tongue and nose, respectively.

All that can flow down and/or is susceptible to taste has the water element. The water element in the juice of a fruit or a vegetable is apparent. A piece of jaggery or sugar candy, though solid, has an element of water in a subtle form because they can be tasted. From this broad perspective, water is present in almost everything we use. It has been the culture of the land to treat every created thing with reverence containing five elements. They are all worshipped.

80 आकाशम्, aakaasham
81 वायुः, vaayuh
82 अग्निः, agnih
83 आपः, aapah
84 पृथ्वी, पृथिवी, pruthvi
85 शब्दः, shabdah
86 स्पर्शः, sparshah
87 रूपम्, roopam
88 रुचि/रसः, ruchih/rasah
89 गन्धः, gandhah

The five elements earth, water, fire, air and ether are perceivable by our sense organs; the converse are those that are perceivable by our sense organs are made up of these five elements. These are inert and are without consciousness.[90] Another entity is the subjective consciousness[91] which depends upon/resides in a body for its manifestation and expression, and the actions and reactions of an inert body which are visible and can be experienced imply the presence of the subjective consciousness within that physical structure. This physical structure is made up of the five elements, water being one.

Ayurveda, the science of life describes water as that which is volatile, cold, heavy, sticky, relatively slow moving and dense,[92] and consuming it imparts to the body and mind a nature of attachment, dampness, rawness, freshness, enthusiasm and binding.

In Indian literature the qualities and the importance attributed to the water resource can be understood through a description of the words used to call it. In the language of Sanskrit, there are at least hundred words used for calling water. Sanskrit language which has its origin in the Vedic language has a very systematic procedure of developing a word. Every word is formed from a root called 'Dhatu' with its meaning 'Dhatvarthah'. In this process, water has been named with a meaning, describing its qualities accorded to it in the language. Some of the words used for water in Sanskrit with their meanings are listed below. The words are taken from Nirukta, the thesaurus of Vedas and Amarakosha, a book of synonyms of Sanskrit language.

Generally, in any language, most of the words get their meaning by the force of usage and convention. But the Vedic language and Sanskrit derive a word scientifically and rationally step by step in accordance with the well laid-down rules of grammar. In this process of derivation, they imbibe different features of the concept or the person or the creature in a codified manner, employing specific prefixes and suffixes; this is unique to Vedic language and Sanskrit.

A deep study of the words derived in the above manner unfolds different concept processes and applications related to, in the present context, the Water resource (Table 1).

---

[90] जड, प्रकृति, अचेतन/ Jada, prakruthi, achethana

[91] जीवात्मा/Jeevaatmaa

[92] द्रवशीतगुरुस्निग्धः मन्द्रा सान्द्र गुणोल्बणं आप्यम् विश्यन्दक्लेदप्रह्लदबन्धक्रुत्.......अष्टाङ्गहृदयम् ९.६ drvasheetagurusnigdhah mandraa saandra gunolbanam aapyam vishyandakledaprahladabandhakruth……..Ashtaangahrudaam….9.6

Table 1 Dileneation of qualities of water as described in Vedic and Sanskrit language

S N	Root/धातु	Root meaning/धात्वर्थः	Meaning of the word/qualities of water
1	अम्भः, Ambah, आपः, Aapah	आप्लृ व्याप्तौ	That which spreads
2	वशिम्, Visham	वशिलृ व्याप्तौ	That which expands
3	अहिः, Ahih	अहि व्याप्तौ	That which pervades
4	कबन्धम्, Kabandham	कमु कान्तौ	That which beholds, which binds the Praanavaayu, oxygen
5	मधु, Madhu	धम गतौ	That which is enjoyable as honey
6	पिप्पलम्, Pippalam	पृ पूरणे	That which quenches thirst
7	क्षपः, Kshapah	षप प्रेरणे	That which quenches thirst and energises
8	पुरीषम्, Purisham	पृ पालनपूरणयोः	That which nourishes plant and animal life, that which fills during Pralaya
9	जन्म, Janma	जनि प्रादुर्भावे	That which was formed in the beginning of the universe and that which gives birth
10	सुक्षेम, Sukshema	क्षी नविसगत्योः	That which settles/establishes all living creatures
11	जामि, Jaami	जन परादुर्भावे	That which is responsible for sprouts, growth and offshoots
12	आयुधानि, Ayudhaani	यू मश्रिणामश्रिणयोः, युधि संप्रहारे	That which supports throughout the span of life of a creature, that which smashes on falling and flooding
13	सहः, Sahah	सह मर्षणे	That which gives strength, that which is strong, that which gives tolerance against heat/temperature
14	क्षोधः, Kshodhah	क्षुदर् सम्पेशणे	That which gets scattered on falling over the boulders while falling down from the mountains
15	घृतम्, Grtham	घृ क्षरण सेचने, दीप्त्योः प्रसवने	That which is sprinkled by clouds, which pours down, that which shines/reflects
16	कशः, Kashah	कश गतौ	That which rushes to a lower level
17	बुसम्, Busam	ष्णा शौचे, बुस वसिर्गे	That which cleanses, that which is renounced/released by the clouds
18	बर्बुरम्, Barburam	पृ पालनपूरणयोः	That which fills and nourishes
19	सीरा, Siraa	सृ गतौ	That which is moving
20	धस्मन्वत्, Dhasmanvat	ध्वंसु गतौ	That which destroys which crosses its path of flow
21	अक्षरम्, Aksharam	क्षर सन्चलने	That which exists in different forms
22	स्रोतः, Srothah	सृ गतौ	That which springs and flows
23	सरः, Sarah, सलिलम्, Salilam	सृ गतौ, सल गतौ	That which flows

(continued)

Table 1 (continued)

S N	Root/धातु	Root meaning/धात्वर्थः	Meaning of the word/qualities of water
24	भॆषजम्, Bheshajam	भषि चकित्सायाम्	That which heals, that which is a medicine or that which is employed in treatment
25	जलम्, Jalam	जल अपवारणे, जल घटने	That which is sharp and that which covers
26	तॊयम्, Toyam	तु सौत्रॊधतुः आवरणे	That which encompass
27	वारि, Vaari	वृङ् वरणे	That which is worthy of selection
28	कमलम्, Kamalam	कमु कान्तौ	That which is desired
29	पयः, Payah,पानीयः, Paaneeyah	पीङ् पाने	That which is worthy of drinking
30	क्षीरः, Ksheerah	घसलृ अदने	That which is potable
31	अमृतम्, Amrutham	मृङ् प्राणतयागे	Absence of which causes death and presence of which prolongs life
32	जीवनम्, Jeevanam	जीव प्राणधारणे	That which sustains life
33	पातः, Paatah	पा रक्षणे	That which protects
34	पुष्करम्, Pushkaram	पुष पुष्टौ	That which nourishes
35	भुवनम्, Bhuvanam	भू सत्तायाम्	That which exists predominantly
36	वनम्, Vanam	वन संभक्तौ	That which is shared, spring
37	उदकम्, Udakam	उन्दी क्लॆदने	That which wets
38	अम्बू, Ambu	अबि,अभि शब्दे	That which drones
39	मेघपुष्पम्, Meghapushpam	पुष्प वकिसने	That which blossoms off a cloud
40	घनरसम्, Ghanarasam	रस आस्वादने	That which is tasty
41	धरुणम्, Dharunam	धृङ् धारणे	That which bares the earth, i.e. the life on earth
42	अरिन्दानि, Arindaani	रा दाने	That which bestows unparalleled happiness given by water

It is evident that water is not just H_2O, but an essential resource with many facets. As we delve deep into the meaning, usage and etymology of the above words, we'll be more than convinced that our ancestors had a holistic view of *water.* When these words are used repeatedly over time, the value of water is perceived by the one who uses these words. This goes a long way to give a stance to the user to diligently use the water resource. This has been finely woven into from the mundane to cultural practices.

Mythological Anecdotes

Fictions are enchanting; they mesmerise and sometimes entertain. These form the reasons for creation of anecdotes and stories. They surround famous personalities or things inevitable. Imagination, exaggeration and unrealistic things galore in them and are devoid of any utilitarian application. However, to illustrate, a few anecdotes related to water are narrated.

The Course of Ganges to Earth (Ramayana)

King Sagara of the Surya Dynasty chose to perform the Ashwamedha yagna, but his royal agents lost track of the sacred horse. Sagara directed his 60,000 sons to trace the sanctified horse. The conceited and mercurial princes ranted and raved from corner to corner of Bharat, blazing jungles and displacing life and chattels to locate the stallion. They lastly got to a calm site where the Sage Kapila was seated in rumination. Next to him was tied the white colt. The infuriated princes slammed Kapila as a crook and striked at him. The Sage then unlocked his eyes; his enormous strength crooked the princes into ruins. Sagara's grandson Anshuman went in search of both the horse and his uncles and converged with Sage Kapila who informed the prince on pleading that the lone way for the souls of the deceased princes to rise to bliss would be through the offering of 'Niravapanjali'[93] with the waters of the holy River Ganga, which flowed only in the paradise.

Bhagiratha's Tapasya[94]: Bhagiratha, a descendent of King Sagara who ruled over the kingdom of Kosala[95]; the kingdom had begun to lose its peace and prosperity on account of natural disasters due to the sins of the thousand princes multiplying in destructive energy, and Bhagiratha found the kingdom not possible at all to rule. So he turned over the empire to his ministers and set out to the Himalayas to do a grueling penance in the severe environment. For 1000 years, he carried out an agonisingly rigorous penance to please Lord Brahma to sanctify his territory with the waters of River Ganga. At the end of the penance, Brahma was pleased and made Bhagiratha aware to propitiate Lord Shiva, as just he was able to break Ganga's plunge. Ganga had a powerful flow and it would have been impossible for anyone to restrain the devastating impact of the event apart from Shiva.

[93] Niravapanjali is a sacred ritual in Hinduism where after the cremation rites, the ashes are ceremonially immersed in holy water by the closest relatives, so that the soul may rise to heaven.

[94] Tapas (tapas, Sanskrit: तपस्) means deep meditation, an effort to achieve self-realisation, sometimes involving solitude, hermitism or asceticism; it is derived from the root word tap (Sanskrit: तप् or ताप) which depending on the context means 'heat' from fire or weather, blaze, burn, shine, penance, pain, suffering or mortification.

[95] A kingdom in ancient India, which is now in the state of Uttar Pradesh.

Bhagiratha took to execute a tapasya (penance) centred on Lord Shiva as outcome of which Shiva came into sight and guaranteed Bhagiratha that he would make Ganga drop on his matted locks. Ganga then surged down from heaven, as Bhagiratha and celestial spectators were horrified of the growl and amount of water approaching. But Lord Shiva came into view under the flow and confined all of Ganga in his matted locks (dreadlocks) just prior to her fall on earth. Ganga arrived to this world because of Bhagiratha's labours, so Ganga is also named 'Bhagirathi'.

Jahnavi, the Daughter of Jahnu

Ganga came down in a destructive flow destroying Jahnu Sage's ashram. This angered Jahnu and took Ganga as 'aaposhanam'.[96] Bhagiratha after that pleaded Jahnu to release Ganga and Jahnu pleased. Hence, Ganga is also called *Jahnavi.* Ganga then ran over the relics of the 60,000 precursors of Bhagiratha and fetch to them moksha, the bliss.[97] Ganga is still deemed to run from Shiva's matted locks or jataa down to earth with a composed might.

River Cauvery

There are several legends about how the River Cauvery had her birth. In the Skanda Purana, Chaps. 11–14 narrate many of them. According to the most renowned account, when the grand sea was churned by the gods and the evils, in order to acquire Amrita, the elixir of life, Lord Vishnu transformed himself into Mohini, a non-peril of immeasurable prettiness, to avert the demons and bring back the elixir to the gods. Goddess Lakshmi along with Lopamudra were sent successively. Lopamudra was sent as an apsara to lend support to Mohini. After the elixir was fruitfully restored to the Devas, Lopamudra was espoused by Brahma as his inheritor.

Following this a little later, a sage of fame called Kavera implored to Lord Brahma that he sanctify him with an offspring. Brahma was pleased by his devoutness and bestowed him with Lopamudra. Lopamudra was christened as Kaveri after the Sage Kavera assumed to be her father.

Kaveri was extremely ardent that her father has all the pleasure and wealth in life and a consecrated terrain full of noble and blissful natives. So she departed to Brahmagiri and entreated Lord Brahma that she get transformed into a river and run

---

[96] Aaposhanam refers to taking a sip of water in a single palm, generally right palm. The quantity of water will only be covering the palm leaving fingers.

[97] Moksha (Sanskrit: मोक्ष moksha) or mukti (Sanskrit: मुक्ति) is liberation or release. In eschatological sense, moksha is liberation from samsara, the cycle of death and rebirth.

all through the state, dispensing her blessings on the citizens and making the land rich and productive. She also requested that her waters be so divine that all those who took immersed in it might be released of all their transgressions. Brahma approved her both the boons gladly and Kaveri was truly delighted.

But something else was to occur to her shortly. Sage Agastya saw her when she was intensely meditating on the Brahmagiri hill. He fell in love with her and proposed her to wed him. Even though her courage was set on becoming a river, Kaveri couldn't turn down Agastya. But Cauvery assured him that if ever he left her unaccompanied for long, she would encompass to disown him and get to her thoroughfare. Agastya pledged it and kept his word loyally for a short while. But one fine day, he was eventful in a philosophical debate with his followers and did not adhere to keep up to his time. Kaveri waited tolerantly and a little later, she leaped into Agastya's unique sacred reservoir and gushed from it. As the disciples of Agastya witnessed the happenings, they attempted to obstruct her surge. But Kaveri swiftly went subversive, come into sight at Bhaganda Kshetra, flooded on the road to Valambari and at last joined the Bay of Bengal, and since then River Cauvery is adulated as sacrosanct throughout her course of flow.

There is one more fascinating faith dealing with the River Ganges and Cauvery that these two join underground on one occasion in a year, at some stage, in the month of Tula cited in the Hindu almanac, with the purpose of cleansing herself of the contamination caused by the mass of sinners who immerse in her waters all through the year. Kaveri is contemplated to be as holy as the River Ganga, with the similar power to sluice off sins.

Bhagamandala, where Cauvery unites with the rivers of Kummahole, Hemavathi, Lakshmanatirtha, Shimsha, etc., is deemed to be a much sanctified place. Temples built all along the banks here are visited by immeasurable number of devotees.

Similarly, Tungabhadra and Papanasini Rivers which flow across the south Indian states of Karnataka and Andhra Pradesh are held to dispel people of their transgression on plunging in them.

Promoting Conservation and Protection

Conservation, protection and enrichment of all natural resources are the key duties assigned by the ancient Indian/Vedic/Hindu school of thought to every human being.

It has specially designated the job to 'Vysya', one among the four occupational categories Varnas of Brahmana, Kshatriya, Vysya and Shudra as explained by Vedas.[98] These four Varnas or occupational categories demarcated in the Vedic texts

---

[98] ब्राह्मणोऽस्य मुखमासीत् बाहूराजन्यः कृतः ऊरूतदस्ययद्वैश्यः पद्भ्यां शूद्रोऽजायत................
Brahmanaosya mukhamaasith baahraajanyah kuthah oorootadasya yadvaishyah padbhyaam shoodroajaayathah...Rigveda...10.8.90.12

are often misconceived to be castes. But this in reality is a theory proposing the necessity of four different categories of labour that are a definite and inevitable necessity of a given society.

A human society, at any point of time, will need four different kinds of labour: those who acquire knowledge and spread that knowledge, otherwise called as Brahmins; people who can physically protect society from any danger or disaster, the Kshatriyas; those who distribute the natural resources in the society, the Vysyas; and those who can help society with manual and art works, the Shudras. This is a natural division of labour and has just been indicated sociologically by Vedas. It is not a measure taken by Vedas to stratify a society.

The Vysya, in this context, gets a special reference as his basic duty is conservation, protection, enrichment and distribution of natural resources. But this does not mean that people, apart from Vysyas, do not hold any responsibility with respect to the upkeep of natural resources. Veda demands every human to perform 'Deva Yajna' every day. Deva Yajna refers to nothing but conserving natural resources and using the natural resources diligently and enriching them. Deva means the one who gives[99] and yajna means the best task[100]; performing the best task to the 'Deva' or God who has manifested himself through nature is a duty of every person, and we all are bound to do the best to the nature that has given us whatever we want without any expectation in return.

We find in many ancient scriptures like 'Manusmriti', the code of ancient times in India, several guidelines given repeatedly regarding the protection of waterbodies. It was prohibited to answer the call of nature in the vicinity of a river or a waterbody.[101] It was the responsibility of everyone to keep the river and its surroundings clean and pollution free. Manu instructs further stating 'On the land which is ploughed, in the pyre, on a mountain, in a dilapidated temple, on an anthill never urinate or pass stools'.[102] 'In borrows of animals, while walking, whilst standing, on the banks of a river, at the peak of a hill don't ever excrete'.[103] 'Don't ever defecate facing the breeze, blaze, intellectual, sun and water'.[104] 'Spread some pieces of wood, soil, grass or foliage on earth, dressing the head and the higher cadaver with clothes and eliminate wastes noiselessly'.[105]

---

[99] यज्ञो वै श्रेष्ठतमो कर्मः......................निरुक्तम्/yajno vai shreshtatamo karmah......Niruktham

[100] देवो दानात्............................निरुक्तम्/ Devo daanath....Niruktham

[101] Naapsumootram puriisham vaa shtiivanam vaa samutsrujeth amedhyalipthamanyankaa lohitham vaa vishaani…….Manusmrithi…4.56

[102] Na phaalakrishte na jale na chityaam na cha parvathe na jeerna devayathane na valmalike kadaachana…….Manusmrithi…4.46

[103] Na sasathveshu gartheshu na gachchannapi cha sthithah na nadiithiiramaasaadya na cha parvatha……..Manusmrithi….4.47

[104] Vaayvagninivipramaadithyamapah pashyamsthathaiva gaah na kadaachana kurvanthi vishanmootrasya visarjanam………..Manusmrithi….4.48

[105] Thiraskrithyochchareth kaashtaloshtapatra trinaadinaa niyamya prathayo vaancham samvithaangovaguntithah…………….Manusmrithi….4.49

Manu furthermore cautions ‘anybody who purges wastes facing fire, sun, moon, water, scholar, cow and breezing air will stand to lose his merits and goes insane’.[106]

Manu also imposes codes to prohibit the poisoning of waterbodies during wars which kill innocent people and cattle who draw and drink water from them, stating ‘Also through a war, the water sources belonging to the foes should not be contaminated, as water is priceless and is property belonging to everybody’, though it is contracted to be belonging to the monarch of the land.

Vaasthuratnaakara a renowned book of Architecture says, ‘one who renovates wells, tanks, temples gets eight times results than that of constructing new structures’.[107] Further, it also says ‘one who destroys, wells tanks or temples is a wicked person’.[108]

Padma Purana also rules that a person who pollutes waters of ponds, wells or lakes goes to hell.[109]

Vasishtasamhitaa penned by Sage Vasishta also states ‘do not foul brooks, lanes, pastures, parasol and community spaces by wastes’.[110] The reiteration of which is found in Taittariyaaranyaka that ‘no one should eject or pass urine or spit or bathe devoid of any clothes on the body in any water body’.[111]

---

[106] Prathyagnim prathisooryam cha prathisomodakadvijaan, prathigaam, prathivaatham cha pranchyaa nashyathi mehanah…………Manusmrithi……4.50

[107] वापीकूपतडागेषु देवतायतनेषु च जीर्णान्युध्दरते यस्तु पुण्यमष्टगुणं भवेत्.........वास्तुरत्नाकरः/ vaapikoopatadaageshu devataayaaneshu cha jeernaanyudhdharathe yasthu punyamashtagunam bhaveth..........Vasturathnaakarah

[108] वापीकूपतडागानामाराम् सूरवेश्मतामग्र उच्छेदते निराऽशंक स विप्रो म्लेच्छ उच्यते १८३/vaapiikoopathadaagaanaamaaraam sooraveshmathaamagra uchchedathe niraashanka sa mlechcha uch yathe...................183

[109] सुकूपानां तडागानां प्राणानां च परंतप सरसा चैव भैत्तारो नरा निरयगामिनः पद्मपुराणः १९-७-८/ sakoopaanaam tadaagaanaam praanaanaam cha pramthapa sarasaa chaiva bhaiththaaro naraa nirayagaaminah......Padmapuraanah...19.7.8

[110] न नदयां मेहनं कुर्यान्नपथिनच भस्मनि न गोमये न कृष्टेनोप्ते शाडलोपजीव्य चायासु........... वसिष्टसंहिता/na nadyaam mehanam kuryaannapathinacha bhasmaani na gomaye na krushtenopthe shaadalopajeevya chaayaasu...............vasishtasamhithaa

[111] नाप्सु मूत्रपुरीषं कुर्यात्, न निष्टीवेत्, न वा विवसनः स्नायात्..........तैत्तरीयारण्यकः/naapsu mootrapuriisham kuryath na nishtiiveth na vaa vivasanah snaayaath............Thaiththariiyaaranyakh

Religious Abuses and Misconceptions

It is extremely sad to find water is being polluted by misconceptions and mistaken notions in the pretext of religion which advocates strongly to keep waterbodies free of pollution. River Ganga is a classic example for the rumpus caused by Hindus for wrongly perceiving religion. Vedas have proclaimed to burn bodies on death very clearly[112] but this has not been adhered to by many. The dead, decaying and polluting matter is discarded into the waterbodies without hesitation, and no consideration is set to the values advocated by the great men of the past to adore the natural resources. This is very harshly condemned and is severely punishable under all circumstances in the scriptures.

Sustainable Practices

a. *Natural Farming*: All the civilisations of the world flourished along river banks, with agriculture as a primary occupation. But the Indus Valley civilization did not only flourish on the banks of Sindhu River but also developed into the richest economy in the world depending on agriculture till the advent of British in India. Agriculture in ancient India was the epitome of trades being the provider of all the basic raw materials necessary for survival; this principle has been highlighted by Vedic philosophy stating 'Oh humans toil, work on fields, the prosperity that comes in this way is most tenable and safe, so experience the riches that is procured through this course as the whole lot, do not ever succumb to stake'.[113] Agriculture is the most eco-friendly vocation humans can perform even in the present day and for times forward. Only agriculture can supply food and shelter to mankind. Hence, dependence on agriculture is the only sustainable way of life. Conventionally in India natural farming acknowledged as 'Rishi Kheti', the method of farming followed by sages and a method of cultivation that uses least or no application of fertilizers, even natural and water, were in practice; Indian agriculture featured multi-cropping, use of local and natural varieties of seeds, maximum dependence on rain and minimum dependence on artificial irrigation

[112] भस्मान्तं शरीरम्.........यजुर्वेदः 40.15, Bhasmaantam shareeram....Yajurveda...40.15

[113] अक्षैर्मा दीव्यस्व कृषिमित् कृषस्व वित्ते रमस्व बहुमन्यमानः...............ऋक्ü 10.34.13, krushimit krushasva akshairmaa divyasva, vitte ramasva bahumanyamaamaanah..............Rigveda 10.34.13

by tanks and wells. This method of farming demanded an optimal utilisation of local resources, saved a lot of resources and thus was totally sustainable.

The statistical data of agricultural production in India prior to the arrival of Britons show the numbers much higher than what is being achieved today with the application of chemical fertilisers, insecticides, pesticides and sowing of hybrid or genetically modified seeds in the fields. This practice of using chemical fertilisers and hybrid or genetically modified seeds is not only degrading the soils' quality by mineralising them but also demands higher usage of water.

(b) *Common or Community Wells*: Vedas recommend people to maintain common sources of water, common eating habits, equal labour and equal responsibility so that everyone in the society moves ahead together as a wheel moves with numerous spokes stuck to it from the centre.[114] So it was usual to find in the villages and towns of India to have common sources of water such as a village well or lake against the present-day policy of one tube well for every household. This guaranteed sustainable mining of underground water.

(c) *Joint Family System*: Joint family was a striking arrangement made for an efficient utilisation and management of natural capital. The system ensured reduced burden on natural resources with a common living area at the first place. That common living area was passed on to the forthcoming generations again reducing burden on natural resources. Common hearth, the use of good quality materials that provided extremely good durability, was very much stressed upon for the purpose of passing on things to the following generations. The use of granite for constructing houses, good quality teak and high-quality metals that had medicinal properties such as bronze, brass, copper for utensils were a common practice. From a sociocultural dimension as well, joint family or at least extended families with parents, brood and grandparents would give physical, psychological and social security to all the members reducing the number of destitute in the society. Also the system spelt an unwritten law against fragmentation in the society.

[114] समानी प्रपा सहवोन्नभागः समाने योक्त्रे सहवोयुनज्मि सम्यञ्चोऽग्निम् सपर्यतारा नाभिमिवाभितम्...........ऋग्वेद samaanee prapaa sahavoonnabhaagah, samaane yoktre sahavoyunajme samyanchoognim saparyataaraa naabhimivaabhitam...........Rigveda

Awareness Through Belief Systems

Apart from religious mandates, common belief systems also helped maintain social harmony and ecological balance in the past.

It is believed all over India that 'water' should never be refused to be given to anybody when asked; if done so one will take the life of a lizard in the next birth.

There is also a belief, in normal circumstances, a person should not bathe for more than once in a day and should never take a second bath in the noon because a bath at that time implies death in the family.

Making drawings and designs on floors and walls using water is prohibited; these beliefs uphold the value of water and prevent its wastage.

Similarly during festivities, during the celebration of Gauri festival in southern state of Karnataka in particular,[115] to manage water and also time, some of the family members avoid or postpone taking 'Abhyanjana', a head bath, to the next day on the pretext that the goddess Gauri will have headache if taken bath on that day; this seems to have originated in joint family setups.

It is also believed that in the monsoon month of 'Ashada' (usually the months of July and August), no ceremonies or celebrations such as weddings, the thread ceremony or any like must be held. Since the season is monsoon season, it is recommended to concentrate on the fields and agriculture rather than celebrations to make the best use of rains and natural irrigation.

Such belief systems didn't just help maintain economical use of water resource, but also gave respect to the physical labour of the person drawing water from wells, avoiding misuse of water.

Local Awareness Through Proverbs and Sayings

The importance of water is delineated variedly in the mythology and folklore throughout India. In the mother language of India, Sanskrit, to make one aware of the importance of water, a proverb reads, "a place where there is no financier, Vedic laureate, monarch, physician and a waterway"[116] is not fit for survival even for a single day.

Varaahamihira in his treatise Bruhat samhitaa expresses 'Enter a house where flowers are in abundance, arches are aplenty and embellished with pots filled with water……..'[117]

[115] ಬಿದಿಗೆ ನೀರಿಗೆ ಬೀಳಲಾರೆ ತದಿಗೆ ನೀರನ್ನು ತಾಳಲಾರೆ, bidige nerige bilalaare tadige nirannu taalalaare

[116] धनिकः श्रोत्रियो राजा नदी वैद्यस्तु पञ्चमः पञ्च यत्र न विद्यते न तत्र दिवसं वसेत्/ Dhanikah shrithriyo raaja nadii vaidyastu panchamah pancha yatra na vidyathe na thatra divasam vaseth

[117] भूरिपुश्पविकर सतोरणम् तोयपूर्ण कलशोपशोभितं..............बृहत्सम्हिता..५३.१२५/Bhooriipushpa sathoranam thoyapoorna kalashopashobhitham………Brihatsamhitaa……….53.125

A common proverb in the vernacular, Kannada, to explain the role of water in a human's life says 'when someone is in town, should perceptibly fetch water'.[118]

The doctrines of maintaining water table and other intricacies of ecological balance are expressed through another proverb in the same vernacular Kannada to have at least one tree for one household and one forest for one town.[119]

Another proverb highlights the common ownership of water resources saying 'why is the master's order obligatory to obtain water from the lake'.[120]

Water Purification Practices

The treatment of water is one of the major topics in Ayurveda. Purifying water by heating it and by adding herbs, such as lavancha (a kind of grass), tulasi, doorva and various chemicals like limestone (shankha) and alum and various metals, earthen and rock containers and in combination of the said methods are available in the texts in plenty.[121] These methods were extensively employed in routine life and during special occasions. Copper vessel, soap stone ware, earthen pots, wooden ware and a few fossil stones, gold ware, silverware, iron, brass and bronze wares, etc. now could only be seen preserved as antique pieces.

Some other methods of water purification are elucidated in the Sushruta Samhita. Generally water can be purified in seven fashions immersing in the utensil containing water the Kataka fruits, the gomedha gem or other gems, lotus foliage root, marine mosses, or a semiprecious stone and tying a piece of linen around the neck of the utensil for filtration.

There are seven ways of cooling water, such as exposing a water container to currents of air, immersing the container (tied round with a piece of wet cloth) neck deep in a utensil full of water, churning it with a stick, fanning, siphoning it by means of a piece of linen, burying a water container beneath a sand bed, or keeping it hanging in a pendent bracket.

Manusmrithi, which also refers to water purification practices telling all sorts of fluids, should be filtered and consumed; the wooden materials should be pecked and smoothened to clean up.[122] Manu also says figuratively, keep a step carefully encompassed by sight and drink water that is covered by linen.[123]

---

[118] ಊರಿಗೆ ಬಂದವಳು ನೀರಿಗೆ ಬರಲ್ಲವಾ?/Oorige bandavlu niirige barallava?

[119] ಮನೆಗೊಂದು ಮರ ಊರಿಗೊಂದು ವನ/manegondu mara oorigondu vana

[120] ಕೆರೆ ನೀರಿಗೆ ಏತಕ್ಕೆ ದೊಣ್ಣೆನಾಯಕನ ಅಪ್ಪಣೆ/Kere niirige yaathakke donne nayakana appane

[121] सौवर्णे रजते ताम्रे कांस्ये मणिमयेऽपि वा । sauvarne rajathe thaamre kaansye manimayopi vaa Pushpaavathamsam bhaume vaa sugandhi salilam pibeth.......sushrutha sutraani

[122] Dravaanaam sarveshaam shudhdhiraaplavanam smrutham prokshanam samhithaanaam cha daaravaanaamcha takshanam.....Manusmrithi Chapter 5.11.5

[123] Drishtipootham nyaseth paadam vastrapootham pibejjalam.....Manusmrithi

Varahamihira, a renowned astronomer, in his treatise Bruhatsamhita, in as early as 550 AD, presented the methods of 'phytoremediation' or treating water with plants and their extracts and others for purifying contaminated water. Exposure to sunlight and air and immersing heated stones, gems, gold, silver along with using sand and iron also find a mention. The role of these herbs in purifying the biological contamination of water has now been recognised.

India's ancient wisdom stating brassware are best suitable for storing potable water has gained scientific relevance, as such water stored in brass utensils helps suppress several water-borne diseases say microbiologists.

Further the technique of zoo remediation also was very much prevalent to purify water; it included treating water, particularly well water with tortoises and certain varieties of fishes.

Use of Water in Social and Religious Traditions

Hindus, being the worshippers of nature, employ the forces of nature in their life absolutely. Water, being one of them, is used extensively in the very common and occasional social and religious traditions. The holiness attached to water in the culture clearly reflects the ecological awareness percolating into routine practices. We discuss them in further detail.

Sipping of water to keep the throat wet,[124] activating the nervous system in the sense organs[125] and sprinkling of water around the altar[126] have been inevitable parts of any Vedic ritual.

[124] आचमनम् aachamanam
[125] अङ्गस्पर्शम् angasparsham
[126] जलसेचनम् jalsechanam

It is a practice of Hindus to treat any guest in the following five ways: providing water to wash hands,[127] providing water to wash legs[128] and giving water to sip,[129] a seat to sit[130] and food to eat[131]; this is a common custom in the culture. Kathopanishad refers to this custom stating 'a learned guest who visits our dwellings is gleaming similar to fire and to appease him get water[132]'; in other words guests must be first treated with water to cleanse themselves.

In any ritual sprinkling,[133] to cleanse the place where the ceremony is conducted and the utensils used in the rituals with water is a must.

Water is predominantly used in the rites of Pumsavana, strengthening of foetus, Chooda, 'shaving the head' to remove hair acquired from birth and Upanayana, leading a boy or a girl to the teacher for acquiring knowledge and Samavarthana, the ceremony of convocation.

In the rite of strengthening the foetus in the womb, water is poured in a plate and placed on the right lap of the expectant mother indicating passing on of strength to the foetus through her, and the mother to be is then made to say that she has placed the essence of her power in the holy water.[134]

The tonsuring ritual is performed on a baby aged between 12 and 18 months with a barber being asked to perform the task safely. The razor, an indirect reference to the barber, is requested not to cause harm to the child. It is directed to sterilise the shaving instruments with hot water[135] to prevent septic and infectious conditions.

In the ritual of taking a student to a teacher which is generally understood as the sacred thread ceremony, performed on a boy or a girl in the age group of 8–12 years, a procedure is followed symbolically, called 'Jalaanjali',[136] where the mentor pours water into the joined palms of the disciple which the disciple holds for a while and pours it into a plate to symbolise:

(a) The water indicating knowledge he/she shall cleanse himself/herself of bad impressions or imprints,[137] wash them off, and shall be left with good impressions and imprints

---

127 अर्घ्यम् arghyam

128 पाद्यम् paadyam

129 आचमनम् , aachamanam

130 आसनम् aasanam

131 आहारम्/ahaaram

132 वैश्वानरः प्रविशत्यतिथि ब्राह्मणो गृहान्। तद्दतां शान्तिं कुर्वन्ति हर वैवस्वतोदकम्॥... कठोपनिषत्...१.१.७/ Vashvanarah pravishatytithi braahmano gruhaan. Tadrutaam shaanthim kurvanthi hara vaivasvatodakam.Kathopanishad 1.1.7

133 प्रोक्शणम्,prokshanam

134 अप्सु मे सोमोऽब्रवीत्, apsu me soomoobraveet

135 उष्णेन उदकेनेहि/ushnena udakenehi

136 Holding water in a bowl like formation when the two palms are held together

137 दुष्ट संस्काराः, Dushta samskaarah

(b) Shall collect from teacher all worthy knowledge, hold them, assimilate them, own them and shall spread it to others
(c) Shall inherit all the natural resources and wealth of nature, water being a representative,[138] hold them, preserve and protect them and shall pass them on to the posterity as a true trustee

A lesson of this nature when bestowed/blessed upon an adolescent goes a long way in building the personality of the boy or the girl to respect the environment and that is culture.

Similarly, in the convocation ceremony which is performed at the end of the stay in the Gurukula and the abode of learning wherein the graduate will be presented before and inducted into the regular social life to mark the end of stay in a Gurukula, the student is directed to pour down a few pots of water and drench from head to toe. This symbolises that the outgoing student has drenched himself in the ocean of knowledge. It is a holy and divine bath. Then onwards he/she is referred to as one who has taken the bath 'Snataka'.

Water is used symbolically in the act of giving anything to anybody,[139] and in the belief of consoling the dead[140] (consoling property of water), water has a role to play.

In almost all rituals, a pot[141] made of mud or copper or bronze filled with water, adorned with the sacred items of turmeric, vermillion and coconut with mango or beetle leaves, is prepared for veneration.

Water is also used generously and gloriously for various sacred baths,[142] and a special bath will be done in the ceremony of compensating for the mistakes committed[143] and getting rid of sins. Although a sin is committed through one's body, mind being the root of it, the sinful orientation is at the level of soul. The soul is of course supreme and the body and mind are under its control. The soul is the master and the body and mind the tools of expression and experience of the soul. On the contrary, when the soul is not exercising its supremacy over the body and mind or exercises partially, the body and mind work erratically, and the harmony between the operator and the apparatus will be lost. Regular baths keep the body and mind in tune and well geared and obedient to the soul. This gives the soul poise to use the tool of the body in the best way devoid of sins. It's customary in Vedic/Hindu and several other

[138] उपलक्षणम्, upalakshanam

[139] धारा, दानम्/dharaa, daanam

[140] तर्पणम्/tarpanam

[141] कलशम्, कलशपूजा/Kalasham, Kalashapooja

[142] अङ्कुरार्पणम्, अमृतस्नानम्, अवभृतस्नानम्, चक्र स्नानम्/Ankuraarpanam, Amruthasnaanam, Avabhruthasnaanam, Chakrasnaanam

[143] प्रायश्चित्तविधि/Praayashchiththavidhi

societies to take bath during overjoy and depression. A bath taken at such time balances the moods and temperaments, and the person can react to such a situation normally and in a balanced way.

Respect to Rivers in Scriptures and Mythological Descriptions

Rivers have always been an inseparable part of any civilisation. Reverence to rivers is an age-old tradition. In spite of an increasing pollution of rivers, a theoretical reverence is alive in the tradition and orthodoxy. It is believed in Hindu culture that remembering the holy rivers at the junctures of using water for bath or drinking or otherwise will make that water holy. A shloka meaning 'Let the holy rivers Ganges, Yamuna, Godaavari, Saraswathi, Narmada, Sindhu, Cauvery may assemble, flow into the water ' I am using now'[144] is chanted quite often invariably during worship; it is believed that by chanting this, the deities of the holy rivers are invited and invoked in the water that is about to be used. A few believe that chanting of this hymn showcases the cultural and territorial integrity of the diversified Indian culture.

We find in many scriptures holiness being attached to the rivers calling them goddesses[145] (Rig 10.75.7)[146] (6.45.31),[147] the shatapatha brahmana (13.5.4.11 and 13) and Aithareya Brahmana (39.9). In Bhagavadgita (10.31),[148] Krishna identifies paramatma or god with all the rivers.

The River Ganga in particular has hundreds of verses eulogising its greatness and its sanctifying power in the Ramayana and many Puranas such as the Padma, Narada, Agni and Matsya. Maharshi Vyasa describes, in the epic of Mahabharatha, the inviolability of the river Ganges, 'this auspicious and celestial river is accessible

[144] गङ्गे यमुने चैव गोदावरि सरस्वति नर्मदे सिन्धु कावेरि जलेऽस्मिन् सन्निधिं कुरु/gange yamune chaiva Godavari saraswathi, narmade sindhu kaveri jalesmin sannidhim kuru……..a very popular prayer

[145] अम्बितमे नदितमे देवितमे सरस्वति...........ऋग्वेदः/Ambithame nadithame devithame saraswathi…………..Rigveda

[146] रुजीत्येनी ऋष्ती महित्वा परिज्रयांसि भरते रजांसि। अदब्धा सिन्धुरपसामपस्तमास्वा न चित्रा वपुशीव दर्शता॥.....ऋग्वेदः....१०.७५.७/Rujithyeni rishthi mahitvaa parijrayaamsi bharathe rajaamsi | adabdhaa sindhurapasaamapasthamaasthvaa na chitraa vapushiva darshatha||…… Rigveda…10.75.7

[147] अधि बृभुः पणीनां वर्षिष्ठे मूर्धन्नस्थात्। उरुः कक्षोन गाङ्ग्यः॥.....ऋग्वेदः....६.४५.३१/adhi brubhuh praaninaam varshishte moordhannasyaath | uru kakshona gaangyah………. Rigveda…….6.45.31

[148] पवनः पवतामस्मि रामः शस्त्रभृतामहम्। झुषाणां मकरश्चास्मि स्रोतसामस्मि जाह्नवी॥... भगवद्गीता...१०.३१/pavanah pavathaamasmi raamah shastrabhruthaamaham | Jhushaanaam makarashchaasmi srothasaamasmi jaahnavi………..Bhagavadgeetha………10.31

to all times'. The Ganges water has an external value too, so one should certainly take a dip into it.[149] Shankaraachaarya in his 'Gangaastotram' says in praise of River Ganga, 'he who drinks thy water, O mother ganga will verily attain the highest abode. Yama, the God of death, dare not even look at him who is thy devotee'.[150] Similarly, Jagannatha Pandita's 'Gangaalahari' is another popular work dedicated to upholding the significance of River Ganga.

Like the Ganges River, the Sukumari, Kumari, Sitasi, Venika, Mahanadi, Manijala, Chakshusha and Vardhanika rivers were also considered as sacred.

Almost all the well-known rivers of India have been described in the mythological literature. Iconographical works also ascribe to them specific forms and give detailed descriptions of reverence attached to them.

Pic 6 Devotees reciting Gayatri, the guardian of the primordial energies along the bank of River Ganga. Source: http://books.google.co.in/books?id=if5BWWiEhx8C&pg=PA390&lpg=PA390&dq=Hydrology+in+ancient+India+by+the+National+Institute+of+hydrology&source=bl&ots=QTJZDWy3xx&sig=Ddn9jB6wPWcJwpoZfkmrdxBmvzQ&hl=en&sa=X&ei=ilDCUvnEMsq8rAeCooH4Cg&ved=0CGAQ6AEwCQ#v=onepage&q=Hydrology%20in%20ancient%20India%20by%20the%20National%20Institute%20of%20hydrology&f=false

[149] अनुशासनपर्वः/Anushaasanaparvah

[150] तव जलममलं येन पिपीतं परमपदं खलु तेन गृहीतम्। मातर्गङ्गे त्वयि यो भक्तः किल तं द्रष्टुं न यम शक्तः॥/Thajjalam yena pipiitham paramapadam khalu thena gruhiitham. Maathargange tvayi yo bhakthah kila tham drashthum na yama shakthah

Water Festivals

Worshipping nature and celebrating the wealth of natural resources are loftily performed by Hindus throughout the year variedly. Praying to water in particular among Hindus has taken a beautiful shape of celebration in the form of a number of festivals meant for invoking rains, showing gratitude for receiving good rains and taking oath to protect and conserve waterbodies.

> Off late worship has been construed as an offering of flowers and the like to demonstrate our reverence to the divinity existing in the five basic elements. But a true worship lies in the proper use of things made up of five elements. Thus, in the culture of the yore, eco-consciousness was embedded, and pollution of the five elements in any way was shunned and abhorred.[151]

Some of the most prominent festivals where water is adored, worshipped and enjoyed are listed in the table below (Table 2).

Technologies in Water Management

Awareness Regarding Water Cycle and Hydrology

The Vedic texts, which are more than 3000 years old, contain valuable references to the hydrological cycle. They are found scattered in Vedas in various hymns and prayers addressed to various deities. Likewise, even other Sanskrit literature has valuable discourses regarding hydrology.

Rig Veda (10.6.4)[152] highlights the hydrological cycle stating 'the water which gets divided into minute particles due to the heat of sun and are carried by the wind and after the conversion into cloud, it rains repeatedly'. The hymns (I, 27.6; I, 32.8) clarify that all the water that returns to the blue through the means of wind by the temperature of sunrays gets transformed to clouds, and then after the diffusion by sunrays, it rains to get accumulated in rivers, ponds and oceans.[153] The Veda explains

---

[151] पूजा नाम सत्कारः.....निरुक्तम्/pooja naama satkaarah...........Niruktham

[152] शूषेभिर्वृधो जुषाणो अर्कैर्देवा अच्छा रगु पत्वा जिगाति। मन्द्रो होता स जुह्वा यजिष्ठः संमिश्ळो अग्निरा जिघर्ति देवान्॥...ऋग्वेदः...१०.६.४/shushebhirvrudho jushaano arkairdevaa achchaa ragu patvaa jigaathi | mandro hothaa sa juhvaa yajishtah sammishlo agniraa jigharthi devaan......... Rigveda....10.6.4

[153] विभक्तासि चित्रभानो सिन्धोरूर्मा उपाक आ, सद्यो दाशुषे क्षरसि ऋक् १.२७.६/vibhaktaasi chitrabhaano sindhorurma upaaka aa, sadyo daashushe ksharasi..........Rigveda...1.27.6

नदं न भिन्नमम्ुया शयानं मनो रुहाणा अतयिन्यापः, याश्चिद्वृत्रो महिनापर्यतिष्ठत्तासामहिः पत्सुतः शीर्बभूव....ऋक् १.३२.८/nadam na bhinnamamuyaa shayaanam mano ruhaanaa athiyanthyaapah, yaashchidvrutro mahinaaparyathishtaththaasaamahih pathsuthahn shirbabhuva....... Rigveda....1.32.8

Table 2 A list of water festivals celebrated across India (pictures sourced from google)

S N	Name of the festival	Time of celebration	Place of celebration	Process of celebration	
1	Ganga Dashahara	Month of May/June	Uttar Pradesh	**Dashahara** is the tenth day[a] of the bright fortnight of the month Jyeshta according to Hindu calendar usually in the month of May. It is said to be the day on which River Ganga descended to this earth. So a bath in the river on this day especially at the Dashashvamedha Ghat of Kashi (Benaras) is believed to destroy ten kinds of sins so the festival is christened 'Dashahara'	
2	Kumbh Mela	Kumbh Mela is celebrated at different locations depending on the position of the planet of Bṛhaspati (Jupiter) and the sun. When Jupiter and the sun are in the zodiac sign Leo (Simha Rashi), it is held in Trimbakeshwar, Nashik; when the sun is in Aquarius (Kumbh Rashi), it is celebrated at Haridwar; when Jupiter is in Taurus (Vrishabha Rashi) and the sun is in Capricorn (Makar Rashi), Kumbh Mela is celebrated at Prayag; and when Jupiter and the sun are in Scorpio (Vrishchik Rashi), the Mela is celebrated at Ujjain. Each site's celebration dates are calculated in advance according to a special combination of zodiacal positions of Sun, Moon and Jupiter	Haridwar, Allahabad, Nashik and Ujjain	This is mass pilgrimage of Hindus who gather to bathe in the River Ganga in Hardwar, river Yamuna in Allahabad, river Godavari in Nashik and river Shipra in Ujjain. Kumbh Mela happens to be the biggest peaceful and religious assembly of about 100 million people held every third year in Haridwar, Allahabad (Prayaga), Nashik and Ujjain successively. Hardwar and Allahabad host half or Ardh Kumbh Mela in every 6 years. The banks of the River Ganga that flow across Haridwar, Yamuna and, the legendary Saraswati that flows across Allahabad, the Godawari across Nashik and the Shipra across Ujjain are places where this assembly takes place. The pilgrimage runs for about 45 days to get blessed by the drops of elixir that is believed to fall from the sky from the Kumbha (the Pitcher) carried by the gods while the Khsheera Sagara or the ocean of milk was churned	
3	Hariyali Amavasya	August/September	Rajasthan, Uttar Pradesh, Madhya Pradesh and Himachal Pradesh	It is a monsoon festival celebrated on Amavasya or a No Moon Day of the Shravan month according to Hindu calendar, organised at the lake precincts in the states of Rajasthan, UP, MP and HP to mark the beginning of monsoons. It is the time when the whole province turns green and water flows generously in rivers and falls. The female folk adorn in dazzling colours to rejoice Teej, on the third day of Sharavan or the months of June and July. People chiefly venerate Lord Shiva on this day for prosperity and riches and for good farming term	

(continued)

Table 2 (continued)

S N	Name of the festival	Time of celebration	Place of celebration	Process of celebration	
4	Teppotsavam	Month of February/Ratha Sapthami	All over South India	Teppotsava popurlarly is called as float festival in the temples of South India. The famous temples in South India that celebrate it are located in Mylapore, Thiruvallikeni, Thiruvidaimarathur, Tirupathi, Tirumala and Kumbakonam. Every temple celebrates this on a particular day in the year	
				It is the grace of God which helps the human beings to cross the sea of the mortal **world**[b] and reach the heaven; this will be presented in a symbolic form with a grand display of lights and colours	
				Applications: The float is made of number of mud pots of uniform size tied by bamboo poles, along the edges and rows called a paristal. The air space in the inverted pots builds up pressure due to heaviness of the objects mounted on the paristol and push of water beneath the pots The air locked in the pots hold up the float for the length of the festival appropriately	
				The deities are decorated beautifully, placed in decorated, special vehicles called 'Vimanas' surrounded by the priest offering worship, Nadasvara orchestra (traditional orchestra), and men with their instruments could be seen accompanied by Vedic laureates and others Throughout the music of Nadaswara is heard coupled with Vedic chants making the Vimanas or boat look luminous and electrified indicating great rainy season and harvest. It's a way of conveying gratefulness to the gods and broadly to the lake for holding precious water in water-scarce regions	

5	Teerthayatra	No particular time	All over India	**Teerthayatra** is a pilgrimage considered a must at least once in one's lifetime. The mode of transport earlier used was by foot preferably. It is not that other modes of transport were not known or available, but it was deliberate. Even today a pilgrimage is considered sanctified only when it is taken by walk, preferably barefoot. This helps the pilgrim to gradually get accustomed to the climatic changes in addition to a close study of ecology and culture of the land. A detailed paced appreciation of nature's beauty during the pilgrimage is an added advantage. A dip in the waters, a bath and the water falls on the way and a dip in the holy waters in the pilgrim centres are an unforgettable agenda in the pilgrimage. The waters in the pilgrim centres are considered holy because of their medicinal and health-showering properties. Many pilgrim centres like the Bendr **Teertha**[c] gained popularity because of its health-giving properties. These waters were revered and protected from over-exploitation and pollution. But off late callousness towards natural resources in particular water has rendered such spots unholy and on the verge of extinction.[d] Trips to Kailasa Manasasarovara,[e] Talakaveri,[f] Bhagamandala[g] and others are also taken up to assimilate the significance of the role played by waters in life and offering reverence to them	
6	Deepavali	End of October or the beginning of November	All over India, but (water festival in Deepavali is celebrated only in Karnataka state)	**Deepavali** is celebrated for 3 days, and the second day of it is celebrated as water-filling festival[h] in the South Indian state of Karnataka. On this day every vessel in the household will be filled with water that will be used for a special bath on the following day celebrations of 'Balipadyami'. This shows the time-management strategy that used to be followed	

(continued)

Table 2 (continued)

S N	Name of the festival	Time of celebration	Place of celebration	Process of celebration	
7	Tulasankramana	Month of October	Karnataka	River Cauvery, one of the Sapta Sindhus, the seven sacred rivers, takes origin at Talakaveri in the Brahmagiri hills. The river surfaces as a tiny perennial spring from a trench called Kundike. It is believed that every year on day of Tulasankramana Parvathi, the goddess of power appears in the Kundike with the spring of the river, Theerthodbhava, noticeable by the sudden upsurge of water, hence considered very auspicious. This moment is primely celebrated by the people of Kodagu. The sudden surge of water from the trench is regarded as goddess Cauvery appearing to bless the people. The devotees take a bath in the tank to purify their bodies and souls. The water is also carried home by the devotees	
9	Hartalika Teej	Month of August/September	Rajasthan	On this sacred day women hum traditional songs rejoicing the arrival of monsoon	
10	Ganga Mahotsav	Month of November	Uttar Pradesh/ Varanasi	As per popular belief, God descends to the River Ganga to bathe on Dev Deepavali, which is a full moon day, on the 15th day of the month of Kartika/usually November. It is during this time of the year that the Mahotsav of Ganga is celebrated to pay a tribute to this holy river of Hindus	
11	Ganga Dashami	Month of June	Uttar Pradesh	The festival marks the auspice of the descent of the River Ganga from heavens. It is venerated all along the banks of the river from Garhmukteshwar to Allahabad. Devotees cleanse themselves of their sins taking a dip in the holy river on this day alongside offering sandalwood, flowers and milk. Aquatic fauna are also fed with balls made of flour. At Varanasi a special Aarti, known as Ganga Aarti, is done to mark superiority	

12	Minjar Mela	Month of August	Himachal Pradesh	Chamba district of Himachal state hosts this mela or fair in the month of Shravana in Hindu calendar and the second Sunday of it to commemorate the victory of the king of Chamba in 935 AD and to celebrate the harvest of paddy and maize. 'Minjar', meaning a bouquet of paddy plant wrapped in golden silk and red fabric, is offered to begin with. A flag-hoisting ceremony is also conducted at this occasion at Chougan to begin a cultural and social program week. The images of Lord Raghuvira with 200 more deities are taken in a chariot pulled by folks with the ropes of the chariot. 'Kunjari Malhar', folk dance and music, makes part of the celebrations	
				A parade is taken from the Akhand Chandi Palace to Ravi River, to end the festival with offerings made to the river. Also this event is observed to remember King Sahil Verman who changed the route of the river, to facilitate devotees to access the Hari Rai temple	
13	Boat Festival	Month of October	Kerala	The whole of the southern state of Kerala hosts this boat festival as Vallom Kallies during Onam, the harvest festival conducting boat races.	
				The boat festivals in Kerala are popularly known as Vallom Kallies. The snake boat race of Chundan Vallam is the major attraction of the celebrations	
14	Adiperukku	Month of August	Tamil Nadu	Padinettam perukku or Adiperukku is signified by celebrating the rise of water at the basins of all the perennial rivers and lakes of Tamil Nadu at the onset of monsoon specially from Aadi, the 18th day of the solar month (generally first week of August). Padinettu meaning eighteen and Perukku meaning rising. Women folk predominantly observe this festival. Adiperukku is a water ritual conducted to pay tributes to nature	

(continued)

Table 2 (continued)

S N	Name of the festival	Time of celebration	Place of celebration	Process of celebration	
15	Miscellaneous	Mostly October	All over India	In every town and village when a lake or river gets filled or flows over mostly during the time of ‘Navaratri’, people are found to celebrate the occasion offering flowers and other holy items like turmeric, kunkum and little lamps lighted. In other words it is an occasion to celebrate as a festival	

[a]ज्येष्ठशुक्ल दशमि, Jyesta Shukla Dashami
[b]संसार/samsaara
[c]Bendr Theertha (65 km): A scenic spot on the banks of the Seerehole River, 65 km south of Mangalore and 15km from the prosperous training centre of Puttur. Bendr Teertha is the only hot water spring in South India and is located off the main road, between Sulya and Puttur in Dakshina Kannada district. Bendr Theertha is a natural hot water spring (Bendr is hot in the language Tulu) with curative powers, mainly skin ailments owing to the presence of sulphur. Though this hot spring can hardly be compared to the boiling and steaming, hot sulphur springs that dot the Himalayan foothills, the lukewarm waters of Bendr Teertha do have traces of sulphur (http://www.deccanherald.com/content/63410/will-spring-surprise.html).
Heated water can hold more dissolved solids, and warm and especially hot springs also often have a very high mineral content, containing everything from simple calcium to lithium and even radium. Because of both the folklore and the claimed medicinal value some of these springs have, they are often popular tourist destinations (http://www.india-travelite.com/indiancities/mangaloresightseeing.htm).
[d]The unabated digging of bore wells around the area has the most adverse and direct effect on the hot spring of Bendr teertha (http://www.deccanherald.com/content/63410/will-spring-surprise.html).
[e]The place of the origin of the River Ganga at the helm of Himalayas.
[f]The place of origin of Cauveri River.
[g]The place where Cauveri River joins Kannike and Sujyoti rivers
[h]ನೀರುತುಂಬೋ ಹಬ್ಬ/NiiN Niiruthumbo habba

also that water is never stationary and constantly gets evaporated, and due to tininess of particles, the upward movement of them cannot be seen.[154] More so, verses (V, 54, 2 and V, 55, 5)[155] give details about the cloud-bearing breeze as the source of precipitation.[156]

Likewise Yajurveda (10, 19)[157] and Sama Veda (6, 607)[158] clearly speak of the hydrological cycle. The Chandogyopanishad also gives a vivid picture of the hydrological cycle, perhaps the first mention of it with its accuracy: 'the rivers all discharge their waters into the sea.

Moving from one ocean to another, the clouds push them to the sky in the form of vapour and they are again released as rain'.

Numerous Puranas elucidate about several stages of the hydrological cycle. In Linga Purana, a whole chapter (I, 36) (I, 36.38-39) is dedicated to the discipline of hydrology. It scientifically explains evaporation, condensation and rainfall with suitable examples besides explaining that water cannot be destroyed; only its condition is distorted,[159] which means water escapes from all the bodies on earth to the atmosphere on getting vapourised by sunlight to turn clouds to rain.

---

154 अतिष्ठन्तीनामनिवेशनानां काष्ठानां मध्ये निहितं शरीरम्, वृत्रस्य निण्यं विचरन्त्यापो दीर्घं तम आशयदिंद्रशत्रुः (ऋक् १.३२.१०) athishtanthinaamaniveshanaanaam kaashtaanaam madhye nihitham shareeram vrutrasya ninyam vicharanthayaapo deergham thama aashayadindrashtruh……….Rigveda…..1.32.10

155 प्र वो मरुतस्तविषा उदन्यवो वयोवृधो अश्वयुजः परिज्रयः। सं विध्युता दधति वाशति त्रितः स्वरन्त्यापोऽवना परिज्रयः॥...ऋग्वेद....५.५४.२/pra vo maruthasthavishaa udanyavo vayovrudho ashvayujah parijrayah | sam vidhyuthaa dadhathi vaashathi thrithah svaranthyaapovanaa parijrayah ||………..Rigveda….5.54.2

उदीरयता मरुतः समुद्रतो यूयं वृष्ठि वर्षयथा पुरीषणिः। न वो दस्रा उपदसयन्त धेनवः शम्भुं यतामनुरथा अवृत्सत॥....ऋग्वेदः..५.५५.५/ udirayathaa maruthah samudratho yuyam vrushtim varshayathaa purishinah | na vo dasraa upadasyanthi dhenavah shambhu yathaamanurathaa………….. Rigveda….5.55.5

156 प्र वो मरुतस्तविषा उदन्यवो वयोवृधोः अश्वयुजः परिज्रयः, सं विद्युता दधति वाशति त्रितः स्वरंत्यापोऽवना परिज्रयः...ऋक्..१.५४.२/ pra vo maruthasthavishaa udanyavo vayovrudhi ashvayujah parijrayah sam vidyuthaa dadhathi vaashathi thrithah svaranthyaapoovanaa.......Rigveda...1.54.2

उदीरयथा मरुतः समुद्रतो यूयं वृष्टि वर्षयथा पुरीषणिः,न् वो दस्रा उप दसयन्त धेनवः शुभं यातामनुरथा अवृत्सत............ऋक् ५.५५.५/udiirayatha maruthah samdratho yuyam vrushtim varshayathaa puriishinah, na vo dasraa upa dasyanthidhanevah shambhum yaathaamanuratha

157 प्र प्रवर्तस्य वृष्भस्य पृष्ठान्नावश्चरन्ति स्वसिच इयानाःताऽआवृत्रन्नधरागुदक्ताऽअहिं बुध्यमनुरीयमाणाः विशणोर्विक्रमणसि विष्णोर्विक्रान्तमसि विष्णोः क्रान्तमसि....यजुर्वेदः.....१०.१९/pra pravarthasya vrishabhasya prushtaannaavashcharanthi svasicha iyaanaah thaavruthrannadharaagudakthaahim budhyamanuriyamaanaah vishnorvikramanasi vishnorvikranthamasi vishnoh kraanthamasi.....Yajurveda...10.19

158 समन्या यन्त्युपयन्त्यानयाः समानमूर्वन्नद्यस्प्रृणन्ति। तमू शुचिङ् शुचयो दीदिवाङ्समपान्नपातमुपयन्त्यापः॥....सामवेदः..६०७/samanyaa yanthyupayanthyaanayaah samaanamurvannadyasprunanthi | tamu shichin shuchayo didivaanjsamapaanapaathamupayanyaapah............Saamaveda.....608

159 दन्धैमनैषु चरचैषु गोधूंभूतस्त्वभः नष्क्रमन्ती या या ऊर्ध्वा मस्त्रैनैरित वै तस्तस्त्वभम्यग्निवायुच. अतो धूमाग्निवतनं संयोगस्त्वमुच्यते वारीणि वर्षतीत्यभ्रंभरस्येषः सहस्त्रादिक्.........लिङ्ग पुराणम् १.३६/dandhaimanaishu charachaishu godhubhuthasthvabah nashkramanthi yaa yaa urdhvaa mastrainairitha vai thasthasthvabhamyagnivaayucha atho dhoomaagnivathanam samyogasthvamuchyate vaarni varshathityahrambharasyesha sahastraadik…..Linga Puranam…1.36

Thus, a combination of smoke, fire and air is the cause behind cloud formation. These clouds cause rainfall under the guidance of Lord Indra, having 1000 eyes. The Linga Purana also states that water can neither be created nor destroyed; only its state is changed.

In Matsya Purana (1,54.29-34, Sharma, 1989) and Vayu Purana (51.14-16: Shastri, 1987), there are references to the hydrological cycle. Vayu Purana (51. 14-15-16) states like this, "the water evaporated by sun rises to atmosphere by means of the capillarity of air, and gets cooled and condensed there. After the formation of clouds, it rains by the vigor of breeze. In this way, aqua doesn't go missing in any of these manners, but gets changed from one form to the other endlessly".[160]

Agriculture was a multifold activity in India encompassing a wide spectrum of branches. Evidences suggest that during Vedic period, agriculture developed as a science. Some of the treatises have been solely dedicated to agriculture written by Parashara and Surapala. Some of the important things worth noticing in it include details of 'Krushipanchang', particularly the methods of a standardised measurement of rain and forecasting. Alongside a high importance was placed on irrigation.

Several apparatus were developed to measure rainfall during the fourth century BC, the time of Kautilya that indoctrinated the principles of modern hydrology excepting the use of 'Drona', the weight measure in the place of depth measure. Indians had developed the method and instrumental devices for measuring rainfall by this time. Varshaman[161] was the name given for the rain gauge. Kautilya says 'at the door of store in a house, a utensil, Kunda of nearly 18 inches wide mouth could serve as a rain gauge' (Arthasastra, Book II, Chap. V).

Kautilya was well versed with the distribution of rainfall in various areas. He furnishes a very accurate scientific account of it with statistical data.[162] The amount of precipitation received in the Jangala or the desert countries is 16 dronas,[163] amounting to one half of what is received in moist countries, 13.5 dronas in the Asmakas country (the state of Maharashtra at present), 23 dronas in Avanti, the border of the Himalayas and the countries where water channels were used in agriculture. From this, it is evident that the essence of the methodology for measurement of rainfall given by Kautilya was the same as we have today with the only

---

[160] आदित्यपतीतम् सूर्यगणेह सौमं सन्क्रमतै जलम्, नदीभिर्वायुयुक्ताभिर्लोकधनं प्रवर्तितै, यत्सोमत्स्त्रवतै सूर्य तद्भैष्ववतीष्टतै मेघ वायूनिघतै विस्रजन्त जलं भुवि, ऐवमुतिक्शप्यतै चैव पततै चंपुनर्जलम् ना नश्मूदकस्यस्ति तदेव परिवर्ततै/Aadityapatitham suryaganeha somam sakramatai jalam nadibhirvaayuyukthaabhirlokadhanam pravarthitai, yathsomasthsthraivathai surya thadbhaishvavathishtathai meghaa vaayunighatai visrajantha jalam bhuvi evamathikshapyathai chaiva pathathai champunarjalam naa nashmoodakkasyasthi thadeva parivarthatha

[161] The historical records and archaeological findings indicate that rain gauges were installed for the first time in India.

[162] द्रोणाष्टांशाम्यधिके वृष्ठोगर्भत्रुतो भवति २ / Dronaashtaamshaamyadhike vrustogarbhathrutho bhavathi 2 Kautilya

[163] 10

difference being that he expresses it in terms of weight measures, while we use a linear measure nowadays (Arthashastra, Chap. XXIV, Book II, P. 130).

Further, while discussing the geographical details of rainfall, he observes 'if a third of the necessary measure of rain occurs during the beginning and final months of the rainy season, with two thirds in between, the rainfall can be considered very even'. Talking about the categorisation of clouds and the relationship between rainfall and farming, the author states, 'the clouds that continuously shower for seven days; and those which pour tiny drops; and that which come into view along with the sunshine are the various kinds of clouds'.

The treatises of Brurhat Samhita (~550 AD) and Mayuracitraka produced by Varahamihira[164] are two very important works which are abounding with the data of climatology and meteorology. The Brihat Samhita has three chapters (21st, 22nd, and 23rd on climatology and meteorology). Chapter 21 speaks of pregnancy of clouds, chapter 22 speaks about pregnancy of air, and chapter 23 speaks of quantity of rainfall. Slokas 1 and 2 of Dakargelam (Chapter 54 of Vrahat Samhita) state the importance of the science of groundwater exploration which helps man ascertain the existence of water veins beneath the earth that resemble veins in the human body. The rainwater assumes various colours and tastes based on differences in the nature of the earth.[165]

Varahamihira has also referred to water divining[166] art and science where in the natural vegetation and terrains are observed keenly and the presence of the quantity of and also the nature of the underground water is identified. Just a few examples are given hereunder. 'The underground water resembles the veins in the human body'. 'The color and taste of the rain water undergoes changes owing to the difference in the nature of the land upon which it falls'.[167] 'Wherever a purple berry tree[168] along with an anthill is found to the east of it, sugary or sweet water is indicated at a

---

[164] Varāhamihira (505–587 CE), also called Varaha or Mihir, was an Indian astronomer, mathematician, and astrologer who lived in Ujjain. He was born in Avanti region, roughly corresponding to modern-day Malwa to Adityadasa, who was himself an astronomer. According to one of his own works, he was educated at Kapitthaka [1]. He was considered as one of the nine jewels (Navaratnas) in the court of legendary ruler Yashodharman Vikramaditya of Malwa.

[165] धर्म्यं यशस्यं च वदाभ्यतोऽहं दकर्गलं येन जलोपलब्धिः, पुं सां यथाङ्गेषु शिरास्तथैव क्षितावपि प्रोन्नतनिम्नसंस्थाः ऐकेन वर्गेन रसेन चाम्भ्यश्च्युतं नभस्तो वसुधाविशेषात् नानारसत्वं बहुवर्णातां च गतं परोक्ष्यं क्षितितुल्यमेव......बृहत्संहिता ५४.१/२/ Dharmyam yashasyam cha vadaabhyathoham dakargalam yena jalopalabhdhih pum saam yathaangeshu shirasthathaiva kshithaavapi pronnathanimnasamsthaah ekena vargena rasena chaambyashchutham nabhasto vasudhaavisheshaath naanaarasathvam bahuvarnaathaam cha gatham parokshyam kshithithulyameva……… Brihathsamhitha….54.1/2

[166] Water tracking or water finding discovering the presence of underground water, जलशोधनम्/ Jalashodhanam

[167] पुंसां यथाङ्गेषु शिरास्तथैव क्षितावपि प्रोन्नतनिम्नसम्स्थाः| ऐकेन वर्णेन रसेन चाम्भश्च्युतं नभस्तो वसुधाविशेषात्, नानारसत्वं बहुवरणतां च गतम्||.....................बृहत्संहिता.५४.१२/ pumsaam yathaangeshu shirasthathaiva kshithaavapi pronnathanimnasamsthah | Ekena varnena rasena chaambashchyutham nabhasto vasudhaavisheshaath naanaarasathvam bahuvaranathaam cha gatham……. Brihathsamhithaa …54.12

[168] जम्बू/ Nerale in Kannada/Jamun in English

distance of three cubits to the south of the tree and at a depth of ten cubit'.[169] 'A vein of good water to the west in about three cubits and at a depth of 12½ cubits is indicated where we find and Indian fig tree'.[170] It is further confirmed by a white snake at a depth of five cubits and the presence of a stone as dark as collyrium.

'There will be sweet and never failing water at a depth of nearly 11 cubits at a distance of 3 cubits to the south of an indigo tree,[171] an indigo tree with an anthill nearby',[172] 'where bael[173] and Indian fig[174] tree are found together, it can be construed that there will be water at three cubits in the south direction and at a depth of 15 cubits',[175] 'palm tree or a coconut tree if covered with anthills denotes a southerly water vein at a depth of 20 cubits and at a distance of six cubits to the west of the tree'[176] 'when a thorny tree flourishes in the midst of non thorny trees or vice versa, water can be found at a depth of nearly 18 cubits and at a distance of 3 cubits to the west'[177]. Many such valuable references are found in the 54th chapter of the Bruhatsamhitaa, and study of it by itself is worth a serious research.

Jains have made a considerable contribution to the field of meteorology. The 'Prajnapana' and 'Avasyaka Curnis' supply stupendous information of a variety of winds. This tradition must have been far older than these treatises. The 'Prajnapana' makes a reference to snowfall and hailstorm. The 'Trilokasara' of Nemichandra names seven kinds of episodic clouds that shower for 7 days during the rainy season. Then there are 12 species of white clouds. They also bring rains for 7 days each. Thus, the season of rainfall extends over 133 days in all.

---

[169] जम्बू वृक्षस्य प्राग्वलीको यदि भवेत् समीपस्थः| तस्मात् दक्षिण पार्श्वे सलिलं पुरुषद्वये स्यात्॥........बृहत् संहिता ५४.९/jambu vrikshasya praagvaliko yadi bhaveth samipasthah । Thasmaath dakshina paarshve salilam purushadvaye syaath...... Brihathsamhithaa...54.1

[170] पश्चादुदुम्बरस्य त्रिभिरेव करैर्नरद्वयेसार्धे। पुरुषे सितोऽहिरष्मान्जनोपमोऽधः शिरासुजला॥........... बृहत्संहिता..५४.११/pashchaadudumbarasya tribhireva karairdvayesaardhe । purushe sitohirashmaanjanopamodhah........ Brihathsamhithaa...54.11

[171] निर्गुण्डि........Nirgundi

[172] वल्मीकोपचितायां निर्गुण्ड्यां दक्शिणेन कथित करैः पुरुषद्वये सपादे स्वादु जलं भवति चाशोष्यं..................बृहत्संहिता....५४.१४/valmikopachitaayaam nirgundyam dakshinena kathitha karaih prurushadvaye sapaade svaadu jalam bhavathi....... Brihathsamhithaa...54.14

[173] बिल्व/bilva

[174] उदुम्बरा/udumbara

[175] बिल्वोदुम्बर योगे विहायहस्तत्रयन्तु याम्येन पुरषैस्त्रिभिरम्बु भवेत्........बृहत्संहिता....५४.१८/ bilvodumbara yoge vihayastatrayantu yaamyena purushaisthribhirambu bhaveth........ Brihathsamhithaa...54.18

[176] वल्मीक संवृत्तो यदि तालो वा भवति नारिकेला वा। पश्चात् षड्भिर्हस्तैर्नरैचतुर्भिः शिरा याम्या...... बृहत्संहिता...५४.४०/valmika samvriththo yadi taalo vaa bhavathi naarikelaa vaa । pashchaath shadbhirhasthairnaraichaturbhih........ Brihathsamhithaa...54.40

[177] कन्टक्यकण्टकानां व्यत्यासेम्भः त्रिभिः करैः पश्चात् खात्वा पुरुषत्रितयं.......५४.५३/kantakyakantakaanaam vyathyaasembhah thribhih karaih pashchaath khaathvaa purushathritayam..... Brihathsamhithaa...54.43

Buddhists too, at least before 400 BC, attempted at a very scientific classification of clouds; four varieties revealed by them are comparable with the most significant four sorts enumerated in modern meteorology. So much of a subtle observation at such an early date is an achievement of the finest order.

The references to the hydrological cycle make it abundantly clear that the ancient Indians were well aware of the different components of the hydrological cycle, the role of solar energy in sustaining the processes and also the general distribution pattern of water. The definitions of hydrological cycle found in some of the ancient works are as good as the modern definitions on this scientific concept.

The impact of the yajna, woods and tanks on precipitation, cataloging of clouds based on their colour, estimation of rainfall based on naturally visible events as the colour of the sky or clouds and the direction of the wind and lightning and the behaviour of flora and fauna all were well understood in India even before the tenth century BC. An ancient text Kadambini describes such method of forecasting of rains in detail.

Traditional Sustainable Practices

A study of history, say a few decades earlier, unfolds before us the systems of managing and monitoring and coordinating with nature in a very simple and decentralised way. The whole modus operandi was woven into the agrarian and community/social routine way of life.

Prayers are generally understood as a besceechment, entreaty or asking the divine forces for something. But, in fact, prayer means complimentary volition[178] and its implementation. The prayers will not be answered unless suitable steps are taken. So it implies that we, our projects and our way of life, should not intervene with the nature in general and in this context natural water cycle. On the contrary, water cycle should be sustained. So, several practices existed to maintain water cycle, some of which are mentioned below.

The link between forests and rainfall was well understood and interwoven with life and religion and took the shape of 'Devavanam/Devara kaadu'/the temple forests. As recently as five decades ago, every temple in India had its own forest, an area which nobody owned or possessed and exercised rights upon, but today there is a severe destruction of culture by consecutive governments, and the practice is now mostly limited to a place in Karnataka called Kodagu. Panchavati or Panchavalkala forest is a miniature forest of five different trees, bael (Bilvapatra), a detoxifying tree; gooseberry large (Bettada nalli), a good immuniser; caraka (Ashoka), which is a coolant; banyan (Aala) again a coolant; and peepal tree (Ashwatha) used to be a common sight. This miniature forest used to be part of every village or locality. These trees were also part of every temple as well.

[178] प्रार्थन वै सन्कल्पः.......................निरुक्तम्/praarthana vai sankalpah....Niruktham

The above trees are ecologically very significant as they do not just possess medicinal and curative properties, but help maintain water tables as the roots of the trees spread very deep into the ground and upgrade the water-hoarding capacity of the earth under it. Since the branches of the trees spread a lot wider, the evaporation rate of water will also be checked to a great extent.

Apart from Panchavalkala or panchavati forests in several places, 'Nakshatravana', a miniature forest that used to have 27 different trees associated with 27 nakshtras or Stars mentioned in Hindu calendar; 'Raashi Vana', a tiny forest that used to have 12 different trees associated with 12 Rashis or zodiacs mentioned in Hindu calendar; and 'Nagavanas' were also there, which exist even today, but rarely.

Gomala: These were pasture or grazing lands for cattle. These used to be maintained in every village to have a constant supply of fodder for the cattle and also to prevent forested lands from being overgrazed by animals, thus passively helping the conservation of water through forests. These were grown, away from ponds and waterbodies.

Besides maintaining forests, measures were also taken to see that the water seeped into the underground levels that would help it have an upward movement. This was achieved by digging trenches and pits in the catchment area everywhere. Vegetation of all varieties used to be planted and grown which, in association with the roots, would check the flow of water and help in the upward movement of water. Some of the other sustainable practices were small tanks with vents—this was a simple system of Kodi (in vernacular Kannada) which determined the volume of water to be collected in a tank with the surplus water flowing out to fill the next tank. Thus, linking tanks in a chain allowed harnessing maximum amount of rainwater without causing any ill effect.

Pic 7 Small tanks with vent. Source: https://www.google.co.in/search?q=manmade+lakes+with+vents&biw=1777&bih=930&source=lnms&tbm=isch&sa=X&ved=0ahUKEwjJxsPT-ZXNAhWJNY8KHdhiBBQQ_AUIBigB&dpr=0.9#tbm=isch&q=manmade+lakes+overflowing+in+india&imgrc=6Ku15NQmeXLxGM%3A

The top soil, the fertile layers of the land, naturally gets washed out or eroded due to rains and gets collected in the tanks as silt, thereby reducing the water-storing capacity of the tanks. During summer, when the water in the tanks was minimum/nil, de-silting used to be a major activity in villages. This used to coincide with the post-harvest season wherein the agricultural work was minimum/nil. This would be a community activity with at least one member from each household voluntarily participating in de-silting. The silt removed was ploughed back to the field, recharging, re-establishing and rejuvenating its fertility.

The benefits of concerted community activity are many and invaluable and form the basis of ecological awareness woven into the cultural fabric.

Pic 8 Compiled by Authors

Sustainable Extraction of Groundwater

The water from the wells was lifted using man/animal power, and value of physical labour and water was known. Water was drawn to the required extent and was not wasted because it meant wastage of man/animal power. There was the system of a water wheel run by bullocks

Pic 9 Water wheel run by bullocks. Source: https://www.google.co.in/search?q=Images+of+persian+water+wheel&tbm=isch&tbo=u&source=univ&sa=X&ei=dsLPUsHiOIflrAfkuIGQDA&ved=0CCoQsAQ&biw=1024&bih=598#facrc=_&imgdii=_&imgrc=ICp65apO1nQn9M%253A%3BWdY1atIUtC0LhM%3Bhttp%253A%252F%252Fepanchatantra.com%252FPT%252Fpics%252F789.jpg%3Bhttp%253A%252F%252Fepanchatantra.com%252FPT%252Findex.asp%253FNo%253D102%3B622%3B462; https://www.google.co.in/search?q=Images+of+persian+water+wheel&biw=1455&bih=756&source=lnms&tbm=isch&sa=X&ved=0ahUKEwiQkpbvotzPAhWKQo8KHaE_BokQ_AUICCgB#imgrc=2vQksgQX-6MA8M%3A

Pic 10 Water wheel. Source: https://www.google.co.in/search?q=Images+of+persian+water+wheel&tbm=isch&tbo=u&source=univ&sa=X&ei=dsLPUsHiOIflrAfkuIGQDA&ved=0CCoQsAQ&biw=1024&bih=598#facrc=_&imgdii=_&imgrc=ICp65apO1nQn9M%253A%3BWdY1atIUtC0LhM%3Bhttp%253A%252F%252Fepanchatantra.com%252FPT%252Fpics%252F789.jpg%3Bhttp%253A%252F%252Fepanchatantra.com%252FPT%252Findex.asp%253FNo%253D102%3B622%3B462; https://www.google.co.in/search?q=Images+of+persian+water+wheel&biw=1455&bih=756&source=lnms&tbm=isch&sa=X&ved=0ahUKEwiQkpbvotzPAhWKQo8KHaE_BokQ_AUICCgB#imgrc=2vQksgQX-6MA8M%3A

Or a system consisting of pulley, rope and bucket worked by oxen on an inclined plain[179]

'Thoobu' as described in the vernacular Kannada is a sort of simple minor irrigation systems that ensured rationing of water by local management

Pic 11 Water pulley. Source: https://www.google.co.in/search?q=Images+of+persian+water+wheel&tbm=isch&tbo=u&source=univ&sa=X&ei=dsLPUsHiOIflrAfkuIGQDA&ved=0CCoQsAQ&biw=1024&bih=598#q=Image+of+a+system+consisting+of+pulley%2C+rope%2C+bucket+worked+by+oxen+on+an+inclined+plain+to+lift+water&tbm=isch&facrc=_&imgdii=_&imgrc=mCpmHgEre0qFVM%253A%3BRF_5jfLImLvWoM%3Bhttp%253A%252F%252Fwww.payer.de%252Fquellenkunde%252Fquellen0563.jpg%3Bhttp%253A%252F%252Fwww.payer.de%252Fquellenkunde%252Fquellen054.htm%3B393%3B555; https://www.google.co.in/search?q=water+pulleys+in+combination+with+oxen&biw=1777&bih=930&source=lnms&tbm=isch&sa=X&ved=0ahUKEwiQ1u29g5bNAhXCqJQKHTmVCw4Q_AUIBigB&dpr=0.9#imgrc=ZN4kO_gXpdscKM%3A

[179] (Kapali), (kapale), (kapile), (kavile), (kavule) in Kannada, a vernacular language.

Pic 12 Water pulley in combination with oxen. Source: https://www.google.co.in/search?q=water+pulleys+in+combination+with+oxen&biw=1777&bih=930&source=lnms&tbm=isch&sa=X&ved=0ahUKEwiQ1u29g5bNAhXCqJQKHTmVCw4Q_AUIBigB&dpr=0.9#imgrc=ZN4kO_gXpdscKM%3A

The employment of electricity and irrigation pumps made drawing of water from irrigation wells easier and began an era of squandering of and indiscriminate use of water. Easier a thing got, carless one spends it. As a few decades passed the upper water table depleted, at this juncture instead of coming to the upper layers emphasising on rain harvesting and recharging of wells the greedy went downwards and hit the lower water tables, thanks to the bore wells and submersible pumps. In no time tanks and wells dried and bore wells went deeper and deeper. This has resulted in serious water crisis. Stringent laws checking the digging and using of bore wells and emphasis on harvesting rainwater and rejuvenating well culture are the need of the hour. It should be borne in the mind and heart that there is no other alternative. It is also worth to note that a large number of waterbodies on the surface and the upper layers of the earth help radiate substantial amount of humidity that successfully counters the greenhouse effect or global warming.

'Etha' was another system or equipment for lift irrigation employed for a sustainable extraction of groundwater in fields.

Pic 13 Etha. Source: https://www.google.co.in/search?q=Images+of+persian+water+wheel&tbm=isch&tbo=u&source=univ&sa=X&ei=dsLPUsHiOIflrAfkuIGQDA&ved=0CCoQsAQ&biw=1024&bih=598#q=Image+of+system+water+lifting+consisting+of+pulley%2C+rope%2C+bucket+worked+by+oxen+on+an+inclined+plain&tbm=isch&facrc=_&imgdii=_&imgrc=hsuP7i4nFgAwAM%253A%3BRF_5jfLImLvWoM%3Bhttp%253A%252F%252Fwww.payer.de%252Fquellenkunde%252Fquellen0553.jpg%3Bhttp%253A%252F%252Fwww.payer.de%252Fquellenkunde%252Fquellen054.htm%3B693%3B548

Soaking Pits: Every household used to maintain a pit in the premises to let out the waste water generated at home; this system also used to be modified in case of availability of a larger space in the form of a backyard garden. These steps also helped maintain the water table besides maintaining cleanliness in the living spaces. Today the drains which have replaced the soaking pits are the victims of a severe negligence causing water logging, diseases and destruction of waterbodies along with other miseries.

Architectural Designs for Water Storage and Conservation

The knowledge of hydrology was deep rooted in the science of ancient India with extensive constructions made for water resource engineering. A variety of water structures were in place much before the deliberately believed Greek, Roman or other civilisations. Every province of India had unique water harvesting techniques developed to suit the local geographical and cultural condition.

Ancient India gave birth to 'Shilpashaastram',[180] a science which deals with the creation of 'Shilpa' or the artifacts, in a wider sense, engineering. The word Shilpa which is derived from the root Shila samaadahu means anything that pleases the

[180] Shilpashastra or engineering includes many articles (things), machines, innovations, metals and artificial means.

mind, deviating from the popular notion of presuming only the icons or sculptures as shilpa. 'Shilpa samhitaa' is a compilation of rules and procedures related to shilpashastra. 'Sthapati' is one who has a complete knowledge of shilpashastra, an engineer or an architect.[181]

There are about 1000 Shilpasamhitaas available. The most popular of them are 'Vishwakarma' or 'Mayamata', Bhartha's Vimanasutras and Matangavaasthu.

King Bhagiratha was actually the first irrigation engineer of India and probably even the world witnessed; he drew the path of Ganga to his kingdom to facilitate people with the waters of the river. Sage Kashyapa reclaimed the waterlogged land of Kashmir, the land of Kashyapa. Varaahamihira was the first hydrologist to predict the locations of aquifers. Sage Parashar developed astronomical methods to predict rainfall (Parashara Krushi shastram).

The properties of flowing and still water were described by sage Vasishta and sage Bhrugu much before Archimedes. Water mills were developed in India before they were adopted in Persia (अरिहत्त, rope pots, seen in the previous section). Kautilya gave guidelines for construction of dams, canals and wells, prevention of water pollution, well-organised water pricing system, etc. Various references are available in the Vedas elucidating the importance of an efficient water use so as to reduce the intensity of water scarcity and drought-like conditions. The botanical names of trees mentioned in Bruhatsamhitaa, calamas rotang rattan (vetas vet) and dalbergia latifolia sissoo tree (shishampaaa shishava),[182] are employed in phytoremediation. Apart from this, several texts refer to a description of designs and qualities of ponds and tanks, (जलाश्यातक लक्षण कथनम्), arch bridges (प्रणालीसेतु लकषण कथनम्), water forts (जलदुर्ग लक्षण), river forts, etc. (वाहिनीदुर्ग लक्षण).

In Narada shilpa shastra, the three (Jalashastra) techniques for storage of water 'Stambham', distribution of water; 'Sanchetanam', drainage of water; and 'Samharanam'—are mentioned. Kunda, veshanta, palavala, haouda, kaasara, tadaaga, saara, mahaa saaraa, subhadraa and saagara are the ten types of tanks.[183]

In 'Amarakosha', a thesaurus of Sanskrit language, several kinds of waterbodies and the words affiliated to them are identified such as jalanirgamaa for the valve of the tank and the channel; koopakaaha and koopaha wells; vidaarakaaha for spring; naavyam, navigable river; srootaha for the fountain head; nimnam, gabheeram, gambheeram and hradaha for a deep source of water; uttaanam, for shallow water; agaadham and talasparshe for an abyss; jalaashayaha and jalaadharaha for waterbodies; aahaava and nipaanam for small tanks for domestic animals to drink water from; dhuhu, prahihi and udapaanam for a well from which water is drawn; pushkarini and khaatam, a square manmade lake; akhaatam and devakhaatakam for a natural pond; padmaakaraha and thataakaha for a tank containing lotuses; kaasaaraha, sarasi and saraha, a lake with lotuses; veeshantaha palvalam and alpasaraha for a barrage; vaapi

[181] Shilpashastra literature compiled by Sri G. G. Joshi, Nagpur, works of late K. V. Vaze who has deciphered and translated many ancient Sanskrit texts of shilpasamhitaa.

[182] वेतस वेत, शिशंपा शिशव/vethasa vetha, shishampaa shishava

[183] कुंड, वेशन्त, पलवल, हौद, कासर, तडाग, सार, महासार, समुद्रा, सागर/Kunda, Veshantha, Palavala, Hauda, Kaasara, Tadaaga, Saara, Mahaasaara, Samudraa, Saagara

and deerghika for a lake; aadharaha for a channel; nadi and sarith for a river, taranginii, sahivalinii, thatinii, hraadinii, dhunii, srothasvinii, dweepavathii, sravanthii, nimnagaa and aapagaa for a river; kulyaa for a small dug river, etc.

Panini's Ashtadhyayi, Katyayana'a Vartika and Patanjali's Mahabhashyam have used hundreds of words which throw light on agricultural operations and irrigation prevalent in the past.

In the eleventh century AD king Bhoja came up with 'Samaraanganasutradhara' (समराङ्गणसूत्रधारा), combining all the other ancient works of engineering in which he speaks of Krudaartha or Yantradhara (कृदार्थ/यन्त्रधारा), fountain, Vaariyantra (वारियन्त्र), machine used to move water, Paata yantra (पातयन्त्र), water fall machine, Samanaadikaa (समनादिका), the release of water from a higher level, Paatasamuchchraaya (पातसमुच्छ्राय),

Bored columns were used to let down water from a height and to take it back through an upslant called Uchchraya (उच्छ्राय). The treatise also provides a bounty of references about the Variants of samanaadikaa, Samuchcharaya tatva (समुच्छ्राय तत्व), principle of circulation of water based on the underground canal that brought water from a remote source to a tank, bathroom with shower Dhaaragruham (धारागृहम्), Pravarshana (प्रवर्शण), the shower, Pranaala (प्रणाल) the pipe, Jalamagna (जलमग्न), the subaquatic, Jalayantraputrikaa (जलयन्त्रपुत्रिका), the sprays, Svayamvaahaka (स्वयं वाहक) automatic, Sakruth Prerya (सकृत्प्रेर्य), occasionally propelled, Antaritha (अन्तरित) and Alakshya (अलक्ष्य), the principle of action and motor mechanism hidden from public view and Vaahya (वाह्य), a machine to be carried by another.[184]

The Traditional Water Harvesting Systems That Are Still in Use

'One tank, one temple and a grazing land for cattle for a village' was the concept of our ancestors for supporting a sustainable growth. Tanks are one of the oldest designs in irrigation engineering found all over India. Water tanks served purposes such as controlling floods, checking erosion of soil, recharging groundwater and reduction of run-offs. The management of tanks was given to individuals or village communities or temples. The entire tank system was suitable for direct irrigation for agriculture and was easy to manage. Tanks were constructed using stone, cement, mud or a combination of these.

Rajasthan, an area largely covered by the Thar Desert, has had a long and the best tradition of water conservation. For example, the builders of the famous Bundi and Chittorgarh forts harnessed the natural catchments found in the forts formed by undulating hilltops. Rainwater was collected in several ways and water flowing down the hill slopes was also stored in waterbodies. Two pictures below of Bundi fort illustrate it. Picture 14 shows a water path, i.e. rainwater flowing down a hill purified by a 'Jaal' (sieve) between the two structures. It then flows into the waterbody and gets accumulated as seen in the picture.

[184] Yantras or Mechanical Contrivances in Ancient India, Raghavan V, 1952, Indian Institute of Culture, Basavangudi, Bengaluru

Pic 14 Source:http://www.esamskriti.com/essay-chapters/Traditional-Methods-of-Water-Harvesting-and-applicability-2.aspx

Talab/Bandhis

Reservoirs natural or manmade were called by different names in various places. They were called Talabs, in general, and pokhariyan, talai and bandhi at some places. Big lakes were also named sagar and samand. These waterbodies were constructed for irrigation and supply of drinking water.

Pic 15 Source: https://www.google.co.in/search?q=images+of+talabs

Johads were small earthen check dams built to capture and conserve rainwater, thus improving percolation and recharging of groundwater.

Pic 16 Source:http://www.esamskriti.com/essay-chapters/Traditional-Methods-of-Water-Harvesting-and-applicability-2.aspx

Baoris/bers were community wells, found in the state of Rajasthan, that were used mainly for drinking water supply. Most of them were built by 'Banjaras'. They could hold water for a long time because of an almost negligible rate of evaporation.

Pic 17 Source:http://www.esamskriti.com/essay-chapters/Traditional-Methods-of-Water-Harvesting-and-applicability-2.aspx

Jhalaras were artificial deep tanks, found in Rajasthan and Gujarat, essentially meant for community use and for religious rites and not potable purposes. Often rectangular in shape, Jhalaras were stepped on all or three sides. The Jhalaras gathered subterranean seepage of a waterbody located upstream.

Pic 18 Source:http://www.esamskriti.com/essay-chapters/Traditional-Methods-of-Water-harvesting-and-applicability-2.aspx

Water Temples or 'Step Wells'

The most unique construction for harvesting rainwater in the arid regions of the country was the step well. Step wells are also known as water temples. These constructions ensured a sufficient water supply through periods of drought. Some step wells were dug near tanks to have an access to water all through the year. Step wells are also christened Vav, Vavadi, Bawdi, Bawri, Baoli and Bavadi. These structures are found amply in the states of Gujarat and Rajasthan. Adalaj Vav is a very well-known step well located 20 kms away from Ahmedabad. It is built in the form of a temple that ends into a well. This is about six storeys below ground level.

Pic 19 Source: https://www.google.co.in/search?q=Images+of+persian+water+wheel&tbm=isch&tbo=u&source=univ&sa=X&ei=dsLPUsHiOIflrAfkuIGQDA&ved=0CCoQsAQ&biw=1024&bih=598#tbm=isch&q=images+of+adalaj+vav&imgrc=IvWqAhBse2_i9M%3A

Kunds are covered tanks used to be built to deal with drinking water shortages. Usually built using local materials, kunds were most widespread in the western parched areas of Rajasthan and in places where groundwater was inadequate and tasted saline. In such conditions, kunds provided clean and sweet water for drinking. They were also rampant in Gujarat and Uttar Pradesh states.[185]

Pic 20 The picture is of a Kund in the state of Rajasthan that is still in use. Source: http://www.esamskriti.com/essay-chapters/Traditional-Methods-of-Water-harvesting-and-applicability-2.aspx

Tanka, most dwellings in Bikaner, had underground tanks built used for storing water. Circular holes were made in the ground, lined with refined lime, in which rainwater used to get collected.

[185]copyright Centre for Science and Environment or CSE

Temple architecture is a very interesting discipline for study. Although today, it has been a centre of faith and a place of idol worship, it was originally a community centre wherein people used to gather, listen to lectures and music and witness dance performances. Many a time, the programmes would go on for several days. The presence of a central dais,[186] kitchen,[187] dining hall[188] and parking places for vehicles of that day[189] in the plan of a temple proves the said as a fact. Water supply to the community was essential for cooking to ablutions, and hence, either the temples were constructed on the banks of rivers or a pond was an inseparable part of the temple precincts.

The temples in South India have huge tanks or ponds[190] as part of the temple premises. Some temples have more than one tank. These ponds were meant to serve the needs of the temple besides the supply of water to devotees for cleansing themselves before Darshan. These tanks even help enhance water table. The picture below shows a water tank in the Chidambaram temple, followed by the pond of Meenakshi Temple in Madurai, Tamil Nadu, and the pond of Venkateswara Temple in Tirumala, Andhra Pradesh.

Pic 21 Chidambaram Temple, Tamil Nadu. Source: https://www.google.co.in/search?q=Chidambaram+Temple+pond&biw=1777&bih=930&source=lnms&tbm=isch&sa=X&ved=0ahUKEwjI8qmj-JXNAhXHOo8KHUiTBj0Q_AUIBigB&dpr=0.9#imgrc=VSDRhzTvj9o-kM%3A

[186] नवरङ्गशाला, navarangashala

[187] पाकशाला, paakashala

[188] भोजनशाला, bhojanashala

[189] अश्वशाला, गजशाला, ashwashaala, gajashaala

[190] कल्याणि, सरोवर, पुष्करिणी, kalyaani, sarovara, pushkarini

Pic 22 Meenakshi Temple, Tamil Nadu. Source: https://www.google.co.in/search?q=beautiful+pictures+of+Meenakshi+temple,+Madurai&biw=1366&bih=599&source=lnms&tbm=isch&sa=X&ei=WLGQUr3yKMLArAerwIHQDA&ved=0CAkQ_AUoAQ

Pic 23 Venkateswara Temple pond, Andhra Pradesh. Source: http://withfriendship.com/user/fraud/tirumala-tirupati-temple-photos.php

Harvesting of water in fortresses and hillocks/catchment areas was also a common technique, and it was known to the culture of the land since ages. Two significant examples are Chitradurga Fort in Karnataka, where not even a single drop is allowed to escape.

Pic 24 Chitradurga Fort, Karnataka, and the other one are the lake at the foothill of Shravanabelagola, Karnataka. Source: http://elusive42.windforwings.com/2011_05_01_archive.html

Pic 25 Shravanabelagola, Karnataka. Source: http://en.wikipedia.org/wiki/Kingdom_of_Mysore

Pic 26 Ancient canal. Source: http://www.esamskriti.com/essay-chapters/Traditional-Methods-of-Water-Harvesting-and-applicability-3.aspx

Kuls are water channels found in precipitous mountain areas like Himachal Pradesh and Kashmir. These carried water from glaciers to villages. In the muddy terrains, the kul was lined with rocks to prevent clogging.

In Assam region, big ponds were constructed for preserving rainwater. In some places, the Garh was used to channelise river water to agricultural fields. A Garh is like a big canal, where both sides have big and long embankment and the middle side is left open for water to flow.

It is apparent that scores of communities in India have efficiently employed water harvesting methods to meet their water requirements. The sustaining strategies were appropriately based on the idea of land-water-vegetation. Any surplus (food, fodder, etc.) generated at the village level supported the towns and cities. The prosperity of villages could be attributed to the water harvesting structures, which assured supply of water for irrigation especially all through the year. Efficient use of water, lining of canals and construction of dams; essential needs for the construction of good tanks, bank protection methods and spillways; and additional aspects were paid due attention.

Water Management During Various Hindu Civilisations

Indus Valley Civilisation

The first major human settlements in India are believed to have begun in the valley of Sindhu or Indus River around 3000 BC in the northern and Western India. The civilisation was prominent in hydraulic engineering and had many water supply and sanitation devices that were in use and were first of their kind. Among them are the earliest known systems of flush toilets in the world. Many homes had it constructed and were connected to a universal sewerage pipe. Private wells were also commonly found. City walls served as an obstruction to floods.

Pic 27 Rivers and sites of early civilisations of India. Source: http://books.google.co.in/books?id=if5BWWiEhx8C&pg=PA390&lpg=PA390&dq=Hydrology+in+ancient+India+by+the+National+Institute+of+hydrology&source=bl&ots=QTJZDWy3xx&sig=Ddn9jB6wPWcJwpoZfkmrdxBmvzQ&hl=en&sa=X&ei=ilDCUvnEMsq8rAeCooH4Cg&ved=0CGAQ6AEwCQ#v=onepage&q=Hydrology%20in%20ancient%20India%20by%20the%20National%20Institute%20of%20hydrology&f=false

Dholavira, an important site of Indus Valley, contained several tanks for gathering rainwater. 'The system of Harappans of Dholavira was very efficient in conserving, harvesting and storing water. The unique features of Dholavira are the sophisticated water conservation system comprising of channels and reservoirs used for storing freshwater of rains and storing the water diverted from two nearby brooks. This came in the wake of the desert climate of Kutch, where several years could pass without precipitation. A seasonal stream running across the site was dammed at numerous points.

Pic 28 Water reservoir with steps at Dholavira. Source: http://en.wikipedia.org/wiki/Dholavira

Lothal is located close to the village of Saragwala in the Dholka Taluk of Ahmedabad. Dating back to 2400 BCE, the dock of Lothal was the earliest known dock in the world linking the city with the track of the Sabarmati River on the trade route between Harappan cities and the Saurashtra Peninsula. It was a significant and flourishing trade centre at that time.

In the excavation which resumed in 1961, archaeologists discovered trenches sunk in the northern, eastern and western borders of the stack, bringing to light the cove conduits and ravine and connecting the waterfront with the river. The discovery comprises of a mound, a township, a marketplace, and the dock.

In Lothal in Gujarat, Inamgaon in Maharashtra and other places in the north and western parts of India, small bunds were extensively constructed for harvesting rainwater.

Pic 29 An old well in Lothal. Source: http://en.wikipedia.org/wiki/Lothal

Pic 30 The dock, with a canal opening to allow water to flow into the river, thereby maintaining a stable water level. Source: http://en.wikipedia.org/wiki/Lothal

The *Harappans,* the ancient people who lived in the Indus Valley from about 3300 BCE to 1600 BCE, were far ahead of their time, especially in architecture. Although not every Harappan house had a well, they were quite common and comprised one of the most identifiable feature of Harappan urbanism. Gradually, the streets and houses were heightened due to amassing of rubbish. This is why tall wells are commonly seen at Harappa and in the neighbouring areas.

Mohenjo Daro is the most studied settlement of the Indus Valley civilisation featuring the Great Bath, the first of its kind in the primitive period. The cisterns at the centre of a courtyard are rectangular structures. Across the lane to the north of the great bath, there is an eight-block bathroom arranged in two rows, one on either side of a drain. A well that is double ringed is seen in one of the rooms meant for supplying water to the bath which is reached by a flight of steps on the sides. This antique city also had about 700 wells, and almost all houses had one private well. Indus Valley Civilisation: *Mohenjo Daro*

Pic 31 Great Bath, Mohenjo Daro. Source: http://en.wikipedia.org/wiki/Mohenjo-daro

Water Management During Various Dynasties

Vedic Period

Vedas recommend a common source of water, common food habits and an equal burden of labour so that everyone in the society could move ahead together[191] sowing the seeds of community living which is the actual philosophy of modern communism centuries before Karl Marx. Intervention of law and governance though was warranted many a time to exercise a good control over water management.

Progress of unfailing sources of water such as reservoirs, ponds, lakes, irrigation canals and others came to be observed as an indispensable part of quality governance. Monarchs and rulers not only got a variety of waterbodies built but also encouraged village communities and individuals to build these on their own. Wide-ranging decrees were passed for regulating their construction and preservation as also the proper distribution of water.

During the period of Rig Veda (VI, 24.6),[192] the relationship between slope and velocity of flow was known reflecting the understanding about the physics of flow. Atharvana Veda (II, 3.1)[193] states that the rivers having their birth in the snow-clad mountains keep flowing in summer also; all the major rivers flowing from the Himalayas were perennial since they were fed by rains during monsoon and snow melt during summer. Atharvana Veda (2.3.1) also quotes so regarding management of water that 'if brooks, wells, pools and others are appropriately used and looked after, will lessen the intensity of drought and scarcity of water'.

[191] samaaniiprapaa sahavonnabhagah sammane yoktre sahavoyunajmi samyanchognim saparyathaaraa naabhimivaabhitham.....Atharvaveda...3.30.6

[192] वि त्वदापो न पर्वतस्य पृष्ठादुक्थेभिरिन्द्रानयन्त यज्ञैः। तं त्वाभिः सुष्टुतिभिर्वाजयन्त आजिं न जग्मुर्गिर्वाहो अश्वाः....ऋग्वेदः....६.२४.६/vi thvadaapo na parvatasya pushtaadukthoobhirindraanayantha yajnau | Tham thvaabhih sushtuthibhirvaajayantha aajim na jagmurgirvaaho ashvaah….. Rigveda…..6.24.6

[193] आदो यदवदावत्यवत्कमधि पर्वतात्। तत्ते कृणोमि भेषजं सुभेषजं यथाससि॥.....आथर्वणवेदः.....२.३.१/ aado yadavadaavatyavathkamadhi parvataath | Thaththe krunomi bheshajam subheshajam yathaasasi…..Atharvaveda……2.3.1

Ramayana (7000 BC) King Bhagiratha, the ancestor of **Srirama**, was the first irrigation engineer who tailored the path of Ganga to the empire of Kosala, presently Awadh in Uttar Pradesh. His efforts are well known as 'Bhagiratha Prayatna' like the Herculian task in English.

The Era of Mahabharatha (3139 BC)

People in the time of Mahabharata saved water in big tanks. After suffering to get water in the country, King Yavakri constructed many ponds and tanks. Wells were constructed in various places and houses for an easy and adequate supply of water. The King also ordered the digging of wells on the sides of big Rajamargas/prominent roads for the comfort of the travellers.

Rules for preserving purity and sacredness of water (as rivers, lakes, tanks, ponds, etc.) and description of several rivers, lakes, tanks, etc. are found in the epic of Mahabharata.

In the Kingdom of Chedi, lakes in the forests were full of clean waters. The Gandhamardana mountain lakes were neat with clean banks, and its water had good healing power because of the presence of cane trees around it.

Sage Narada asks Yudhishtira if he has built reservoirs for storing rainwater for irrigation purposes, saying agriculture should not depend on rains alone.

The largest inland salt lake of India, the Sambhar Lake is a bowl shaped encircling the Sambhar Lake Town located 96 kms from the city of Jaipur in Rajasthan. Sage Shukracharya was known to have lived in this region. The Shukraniti, written by the sage, prescribes that the king should construct wells, tanks and reservoirs to store water for agriculture.[194] He also states that the land revenue should be set on the basis of the irrigational facilities available on it.[195] If a farmer has got wells and water canals constructed on his own, he should be exempted from land taxes till the income is double the cost of cultivation.[196] The sage also uses words referring to artificial lake (कृत्रिम नदी),[197] and there is also a mention of a machine called Udayanayantram (उदयनयन्त्रम्) which probably was something like bucket wheel.

---

194 shukraniithi 4.363

195 शुक्रनीति ४ २२४/shukraniithi 4.224

196 शुक्रनीति ४ २३२,२३३/shukraniithi 4.232 & 233

197 शुक्रनीति ४ २३२/shukraniithi 4.232

Mauryan Empire (322 to 185 BCE)

Chanakya or Kautilya, the kingmaker, laid the foundation of Mauryan Empire and penned 'Arthashastra', a book of political economy of ancient India. Arthashastra gives an extensive account of dams and bunds that were built for irrigation during the reign of the Mauryas. The water supply systems were well organised in a framework of stringent policies. Various types of levis were collected from the farmers based on the nature of irrigation. The rate of tax was 25% of the production in case the water was drawn from natural sources such as rivers, lakes and springs. If water was obtained from storages constructed by the Monarch, the rate of tax differed with the method of drawing; it was 20% of the production for water taken manually, 25% for water pulled by bullocks and 33% for that diverted by means of channels. Taxes were exempted if a farmer built or improved irrigation facilities himself. The exemption period was 5 years for new tanks and bunds, 4 years for restoration of old structures and 3 years for clearing overgown weeds along the constructed irrigation amenities.

Kautilya states of all the irrigation works perennial flow is the best and among the perennial projects, that which can irrigate a vast area is most valuable (Chapter 52.3). Arthashastram (2.1.20) reads 'reservoirs should be built using natural springs and water brought from other places'. 'Shades, courtyards, latrines, fire places, places for pounding grains and open spaces are to be used as common properties' Arthashastram (3.8.28).

Chapter 8 of Arthashastra gives an extensive account of laws pertaining to buildings and their architecture.[198] This chapter reads laws that were enacted to design discharge mechanism of waste and waste water from every house, and from each house a water track of adequate gradient at a distance of 3 padas or 1.5 aratnis from the adjacent spot shall be built so that water flows from it in a constant line or falls into the drain. Violators of this shall be penalised with a fine of 54 panas. If a ditch, strides, water track, ladder, manure hill or anything else of a dwelling bothers outsiders hampers others' delight or causes water to get collected and thus damage the wall of an adjoining house, the violator shall be penalised with a fine of 12 panas. If the displeasure is owing to faeces and urine, the fine shall be doubled. The same fine should be meted out also to a tenant who, even when asked to evacuate, lives in the house, along with the owner.

It is also seen from the ancient treatise on administration that a specific officer was responsible for irrigation works. Waterbodies like reservoirs, bunds and tanks were also privately owned, and the owner was free to sell or mortgage them in expectation for a share in the production. Waterbodies had to be maintained by the inhabitants of the village in the absence of the owner.

---

[198] Vastushastra is the name given to the discipline known as architectural science in modern times.

The first version of the water law is also found in this ancient document. It talks of a set of sentences lay down for breach of water laws like:

- Damage other's ploughed or sown field by letting water flood from a tank or a reservoir.
- Causing damage to gardens, parks and bunds.
- Owner of a higher tank preventing the filling of a lower tank
- Failure to maintain the waterbody.
- Out-of-turn drawing of water from a tank.
- Building a well or a tank on someone else's land.
- Selling or mortgaging a waterbody meant for charitable purposes.
- Penalty of death was ordered for breaking a reservoir filled with water.

Development During the First Century BC

Satvahanas who ruled between first century BC and second century AD in southern India pioneered the concept of brick and ring wells. Lake and well irrigation were developed at a large scale in the reigns of Pandyas, Cheras and Cholas also in south India between the first and third century AD, and large constructions were made across the rivers of Cauvery and Vaigai. The renowned Cauvery anicut was a work of this epoch. Tataka, tanks for tapping rain water, were also built on a large scale. Natural depressions were also developed into irrigation tanks.

Pic 32 This oldest dam in the world, Grand Anicut (Kallanai), is located in Tamil Nadu built by the Chola King Karikalan in the first century AD. Source: http://www.tamilspider.com/resources/3363-Grand-Anicut-Kallanai-Tamilnadu.aspx

The Gupta Empire (320 to 550 AD): The Golden Age of Gupta reign was marked by colossal progress in every aspect of civilisation that shaped up the elements of what is in general branded as Hindu culture. Water resources development on a large scale took place during this period.

Drought was common in the Gupta Empire, so irrigation was essential and taxes were levied on water. Every month people gave a day of labour to maintain wells, irrigation, ditches, reservoirs and dams. The province had a very profitable trade through the Mediterranean Sea.

The Bruhatsamhita (54.118; Achyntananda Jha, 1988), attributed to 550 AD, the period of Gupta Reign, describes the utility of ponds for effective storage of water: 'A pond constructed facing west from east holds water for long while that built from north to south gets spoilt by the waves raised by winds'. The shores of such tanks must be protected by planting of certain trees, the species of which are spelt out; one is tempted to compare this with the catchment treatment measures recommended in modern times for soil and water conservation.

Irrigation in Karnataka was properly taken care of right from the Ganga Dynastic rulers' times in the ninth century with the **'Bittuvatta' system** (the tradition of setting some piece of land in every village as 'Bittuvatta'. Persons enjoying such land were charged with the responsibility of the upkeep of the village's irrigation facility) for their proper upkeep.

The Rajput Period (Ninth to Twelfth Century AD)

The Rajput dynasty (1000–1200 A.D.) highly promoted works in the area of irrigation in northern India which is very much evident in the state of Rajasthan which still have innumerable canals, dams, and reservoirs built in their time still in use. Rajputs also got built the beautiful temples at Khajuraho and also many well-known temples in Gujarat and Western Rajasthan in the tenth and eleventh centuries. Their architecture represents a pleasing blend of Hindu and Muslim styles. Among the more notable are forts at Chitor, Gwalior and Jodhpur and the 'Palace of Winds' (Hawa Mahal) in Jaipur. Jai Singh II, the King of Jaipur, got astronomical observatories constructed in Jaipur and Delhi in the early eighteenth century.

Pic 33 Ana Sagar lake. Source: https://www.google.co.in/search?q=ana+sagar+lake&biw=1600&bih=828&tbm=isch&imgil=RCqVxCFEcN2BdM%253A%253BtGx8tV2eZ9PzlM%253Bhttp%25253A%25252F%25252Fwww.columbia.edu%25252Fitc%25252Fmealac%25252Fpritchett%25252F00routesdata%25252F1100_1199%25252Fmuinuddin%25252Fanasagar%25252Fanasagar.html&source=iu&pf=m&fir=RCqVxCFEcN2BdM%253A%252CtGx8tV2eZ9PzlM%252C_&usg=__q9ygqJAq5Xe9_9euhUrmWpjualk%3D&ved=0ahUKEwjH-M_q8ZLNAhUBQY8KHcjFD6sQyjcIMA&ei=uCdVV8e8EoGCvQTIi7_YCg#imgrc=RCqVxCFEcN2BdM%3A

Ana Sagar is a manmade lake sited in the Ajmer city of the state of Rajasthan. The grandfather of Prithvi Raj Chauhan, Anaji Chauhan, got it built in 1135–1150 AD and was named after him. Local populace built the catchments. Jehangir got the Daulat Bagh Gardens made along its banks, and Shahjahan got the Baradari or the pavilions built for it. The lake covers an area of over 13 square kms.

Pic 34 The Mansagar lake. Source: https://www.google.co.in/search?q=man+sagar+lake&biw=1600&bih=828&tbm=isch&imgil=RCqVxCFEcN2BdM%253A%253BtGx8tV2eZ9PzlM%253Bhttp%25253A%25252F%25252Fwww.columbia.edu%25252Fitc%25252Fmealac%25252Fpritchett%25252F00routesdata%25252F1100_1199%25252Fmuinuddin%25252Fanasagar%25252Fanasagar.html&source=iu&pf=m&fir=RCqVxCFEcN2BdM%253A%252CtGx8tV2eZ9PzlM%252C_&usg=__q9ygqJAq5Xe9_9euhUrmWpjualk%3D&ved=0ahUKEwjH-M_q8ZLNAhUBQY8KHcjFD6sQyjcIMA&ei=uCdVV8e8EoGCvQTIi7_YCg#imgrc=RmCjLZZWr_bd7M%3A

During 1596 AD, when there was a severe famine in the Rajput province, there was a consequent acute shortage of water. The monarch of Amer was then motivated to have a dam built for collecting water to beat the severe adversities caused by the crisis. A dam named Man Sagar was built, across the eastern valley between Amer hills and Amagarh hills with earth and quartzite. The dam is about 300 m long and 28.5–34.5 m wide at present. Three sluice gates are provided for it for release of water to the downstream area. The dam and the palace in its centre have gone through renovation under various rulers numerous times, but the last renovation was taken up by Jai Singh II of Amer in the eighteenth century. The king also got numerous other constructions made of historical and religious significance, some of which being the forts of Amer, Jaigarh, Nahargarh, Khilangarh and Kanak Vrindavan Valley, which lie in the vicinity.

Pic 35 Balsamand Lake. Source: https://www.google.co.in/search?q=images+of+Balsamand+Lake&tbm=isch&tbo=u&source=univ&sa=X&ei=8LWmUrHhJ4exrgeEg4CwCw&ved=0CCoQsAQ&biw=1366&bih=599#tbm=isch&q=balasmand+lake&imgrc=MTJGRzM7r3TqOM%3A

Balsamand Lake is located 5 km from Jodhpur, Rajasthan along the road of Jodhpur Mandore Road. The lake was built by Balak Rao Parihar in the year 1159 AD. It was planned to fulfil the water needs of the people of Mandore. The lake stretches to a length of one km and a breadth of 50 m and is 15 m deep.

Pic 36 Rajsamand Lake. Source: https://www.google.co.in/search?q=images+of+Rajsamand+Lake&tbm=isch&tbo=u&source=univ&sa=X&ei=8LWmUrHhJ4exrgeEg4CwCw&ved=0CCoQsAQ&biw=1366&bih=599#tbm=isch&q=Rajsamand+Lake&imgrc=i5_5-CfIlQXBqM%3A

Rajsamand Lake is a lake situated near the town of Rajsamand in the state of Rajasthan. Located at 66 km off from the north of Udaipur, the lake lies between the cities of Rajnagar and Kankroli. The lake is also known by the name of Rajsamudra Lake in Rajasthan. Built by Maharana Raj Singh in 1660, the width of the lake is about 2.82 kms, the length is about 6.4 kms, and its depth is 18 m. The lake has a catchment area of close to 196 sqm as it is built across the rivers of Gomati, Kelwa and Tali. Rajsamand Lake is one of the five popular lakes of Mewar. This lake also has the pride of a magnificent dam built in seventeenth century. A huge embankment is built in white marble on the southern end of the lake. The marble terraces and stone steps touch the waters of the lake. Here, the five toranas (weighing arches) are also found, where Maharana Raj Singh and his descendants arranged Tuladan in which Kings used to weigh themselves in gold before distributing it among the Brahmans.

In eastern India Pal and Sen Kings (760–1100 A.D.) got a number of large tanks and lakes constructed in their empires. Kalhana's Rajtarangini gives a thorough description of the development of irrigation systems in Kashmir in the twelfth century.

The Vijayanagara Empire (Fourteenth to Sixteenth Century AD)

The rule of Krishnadevaraya witnessed several diversion works across Tungabhadra River with the construction of numerous canals of extensive irrigation and water supply to the palaces of the kings and the capital city that was thickly populated. The royal centre of the palace consisted of bath structures and temple tanks. These systems continue to supply water to agricultural fields even to this day, with some changes from time to time. The important channels still functioning are the Raya and the Basavanna canals supplying water to the fields of Hospet and Kamalapuram near Tungabhadra River. The canals were constructed during the medieval period. Currently, the modern Tungabhadra Dam at Mamallapuram feeds the canals as the original diversion structures remain submerged by the Tungabhadra reservoir. The low-lying canal of the Tungabhadra Dam gets integrated with the old Raya canal. The Basavanna canal had originally irrigated large tracts of agricultural land between Vallabhi Puram and Amaravati (Bhavanishankar, 2007).

Madhava Manthri, a Vijayanagara officer at Talakad, raised the famous Madhava Manthri Katte between the places Hemmige and Muduktore across the River Cauvery in about the year 1341, and a record of the year 1638 speaks of it as Madarasa Vodeyara Katte Kaluve. There is an old anicut at Dhanagere across Cauvery in Kollegala Taluk and another at Ganiganur across Suvarnavathi River (Bhavanishankar, 2007).

Around Hampi, the capital of the empire, are the remains of ancient aqueducts and canals as seen in the picture below and were used for bringing water from the Tungabhadra River to the tanks and baths. Water inside the temples was usually supplied by underground aqueducts.

Pic 37 Source: https://www.google.co.in/search?q=Aqueducts+of+Vijayanagara+empire&tbm=isch&tbo=u&source=univ&sa=X&ei=sq6lUu_7MciJrQf2sYHYBQ&ved=0CDwQsAQ&biw=1366&bih=599

Pic 38 Source: https://www.google.co.in/search?q=temple+ponds+in+the+vijayanagara+empire&biw=1600&bih=828&source=lnms&tbm=isch&sa=X&ved=0ahUKEwjwmrWa9ZLNAhWLo48KHbjmC90Q_AUIBigB

Pic 39 A canal linking to a small tank at Hampi. Source: https://www.google.co.in/search?q=A+canal+linking+to+a+small+tank+at+Hampi&biw=1600&bih=828&source=lnms&tbm=isch&sa=X&ved=0ahUKEwj9qsO79ZLNAhUHQI8KHZnNCBUQ_AUIBigB#imgrc=vMl8sb0kYzwfYM%3A

The Wodeyars (1399–1947 AD)

The Wodeyars who ruled the princely state of Mysore from the fifteenth century onwards built a number of irrigation diversion works across the River Cauvery in South India, and many structures are in existence even today working very well. The Bangara Doddi Kaluve aqueduct, carrying a distributary across the River Cauvery, more than four centuries old, is in good working condition and probably the oldest aqueduct in service in the world.

There are many epigraphical records that indicate that irrigation had been practised from historical times in the erstwhile Mysore State, now merged in Karnataka. The inscriptions of the Ganga, Cholas, Hoysalas and the Vijayanagara and Mysore rulers not only speak of tanks, canals, sluices and other means helpful to facilitate irrigation but also mention about the steps taken for the conservation of such means of irrigation like repair of embankment, desilting of tanks, etc.

The creation of a canal by directing the flow of the Lakshmana Tirtha River, a tributary of Cauvery, is mentioned in a record of the year 1669 with the tank named 'Kanteerava Samudra'. Tipu Sultan, the popular Muslim ruler of Mysore State in those times, raised a bund of 70 ft high in Anandur in Mysore Taluk. In the days of Dewan Purnaiah, a renowned minister during the rule of Krishnaraja Wadiyar III, the Sagarkatte Dam, was raised across Lakshmana Tirtha. At Devanur, the construction of a tank, Devarakatte, is spoken of during the nineteenth century. Chunchanakatte has an old dam ascribed to one Chuncha. The Mysore rulers of those times, Kanthirava and Chikka Devaraja, had undertaken a number of irrigation works.

The Mysore rulers of the recent times have constructed the major irrigation works of the present Karnataka state, the most notable of them being *The Krishna Raja Sagara Dam* (in the picture) built across Cauveri River during the reign of Krishnaraja Wodeyar IV as a life-giving river for the Mysore and Mandya Districts, in 1924. Sir Mokshagundam Visvesvaraya served as the chief engineer during the construction of this dam.

Pic 40 Source: http://en.wikipedia.org/wiki/Krishna_Raja_Sagara

Vani Vilasa Sagara Dam was also built in pre-independence time across the river Vedavathi.

The dam is a beautiful piece of architecture, a marvel of engineering, providing water to lots of surrounding cities, towns and villages, which are largely dry lands belonging to the Deccan area of Central Karnataka.

Pic 41 Source: http://en.wikipedia.org/wiki/Vani_Vilasa_Sagara

Budikote Dam, located 125 km off from Bengaluru, was one of the popular dams in the late 1980s, serving eight villages close by.

Pic 42 Source: http://www.google.co.in/imgres?imgurl=http://i.ytimg.com/vi/reTWHCG8Pic/0.jpg&imgrefurl=http://www.digplanet.com/wiki/Kolar_district&h=360&w=480&sz=21&tbnid=YLkQI3LexNjHnM:&tbnh=127&tbnw=169&zoom=1&usg=__azbiiSqHDYE1JLULpZG25ki0nmk=&docid=cOiVELIfDqBYRM&sa=X&ei=WBDVUsOWJsW5rgfKq4D4Aw&ved=0CC0Q9QEwAA

Kempe Gowda who ruled Bengaluru and its neighbouring areas in the state of Karnataka in the fifteenth century had a number of tanks built. The most well known among those were the Kempambudhi, Dharmambudhi, Sampangi and Siddakatte Kere tanks.

Role of Communities and Individuals: In yester centuries, the village communities and individuals were encouraged to build their own water harvesting structures to meet their domestic water requirements. The communities being closely knit had a strong culture of providing voluntary labour and material contributions towards building these facilities for the welfare of all. The social norms of the community members helped in maintaining these facilities, conserving and protecting water from pollution and guaranteeing just distribution. It is discovered that there was no problem of water scarcity wherever the communities were strong, and people on their own built water harvesting structures. On the contrary, situations were bad where the people depended entirely on the state for water.

Some Observations

Water is inert, without consciousness, and is indifferent to an expression of our gratitude or ingratitude, but the traditional perspective is that water is divine and is invaluable. In our ancient culture, we look at all the natural resources and forces of nature as 'Deva' or 'Devatha'. These words never mean god or goddess as construed today. Deva is a Vedic word, meaning 'that which gives'. Once we receive something, we are indebted to the Deva or Devatha,[199] and a feeling of indebtedness and preparedness to clear the debt is termed as 'gratitude'.[200] If it is a person who has given, it is reciprocated by giving something similar or something else immediately or at times of need. If it is nature which has given, a sense of gratitude is defined as:

a) Using it diligently and deliberately avoiding misuse/abuse of it, that is, preventing pollution at any scale
b) Protecting it to be benefitted time and again in return
c) Sharing the benefits with fellow beings and creatures and perpetuating it for the benefit of generations to come
d) To enrich it wherever and whenever possible

It is apparent that a person with a sense of gratitude is aware of the importance and value of the thing or person towards which he/she expresses gratitude. Alternatively a person without gratitude will squander the thing that has bestowed benefits on him and shows disrespect towards a person who has helped. This is what is happening with respect to water and other natural resources in the present context.

Losing this gratitude in the present-day context has been the main cause behind all the disasters and chaos in relation to water in particular and natural resources in general. But, gratitude, in fact, liberates the one who expresses gratitude. This is one of the foundation stones of Indian/Human/Vedic culture. Thus, an expression of gratitude makes one deserve and enjoy perennial fruits. The ungrateful will miss all this for oneself and put others also on tenterhooks. This is the present scenario.

An in-depth analysis of the problematic scenario and a sincere implementation of the remedial measures are the need of the hour, and this is a verbal sketch of the above. This is guided by an awareness of ecology and scientific exposition, an inseparable part of Indian culture.

Finally, it is the duty of every human to identify the purpose of this mortal life and mortal world which is a beautiful path that leads the soul towards the divine bliss in the super soul which could never be achieved causing wreckage in this world. Submission to God is obvious to attain bliss, but submission to God means not just believing in the existence of almighty, but also respecting his creations and leading life in coherence with the laws of the nature.

This is the story of our glorious tradition of water harvesting in short reflecting the wisdom of our ancestors who thought harvesting of water and managing it effectively is an essential element of culture and life with community. This meant that these systems were perceived by the common man as his sacred duty as also the communities as part of a good local self-governance and social responsibility. This ancient knowledge at every level of society ensured an equitable distribution of

---

[199] देवो दानात् .... निरक्तम्, Devoo daanaath…..Nirukttham

[200] देवऋणम्...............Deva rinam

water, which, in turn, formed the basis for an all-round development and prosperity. Let us revive and extend this old knowledge for the advantage of all our people especially those in the rural areas. We can do it. We, the authors, humbly submit that the subject has not been covered exhaustively. There are many other facets to the subject and we have tried our best just to give a bird's eye view of the subject. The final message could be *Thank You Water*, we shall preserve you.

Acknowledgements Amarakosha, twelve.vaarivarga.259,260,261 (published by Prasaaranga Mysore university Mysore, Edited by Louis Rice)

Kittel, Rev F, Kannada English dictionary (Asian Educational Services, New Delhi)

Edited by Baale, Chandrashekara ಕೃಷಿ ಭಾರತ್, (Krushi Bharath), Kannada Monthly

All India Chinmaya Yuva Kendra, August 2003, Awakening Indians about India

Bhat, Dr K Narayana, 1998, Yaskapranitam Niruktham

Chaturvedi, Sudhakara, 2003, Arya Samaj, Bengaluru, Vedokta Jeevana Patha

Ramaswamy Vanitha, 1993, Shree Prakashana, Prachina Vanijya Charitham

Danino Michel, 2009, Landmarks of Science in early India

National Institute of Hydrology, 1990, Dehradun, Hydrology in Ancient India

Navaratna, Shesha, 1994, Samaja Pustakalaya, Dharawad, Manusmrithi

Raghavan V, 1952, Indian Institute of Culture, Bengaluru, Yantras or Mechanical Contrivances in Ancient India

Ramakrishna Math, 1993, Bengaluru, Saswara Rigveda & Taittareya Samhitaa

Vedabhashya Prakashana Samithi, 2001, Arya Samaja, Bengaluru, Saama Veda Samhitaa,

Shastry, GVV, 2011 February, Prakruthi Prathisthana, Tumkur, ರೋಗದ ಭಯವೇಕೆ? (Rogada Bhayaveke)

Somayaji. CS, 2011 Ayurveda Academy, Aushdham Jahnavi Toyam

Appendix

S N	Subject	Popular texts
1	Medicine/Ayurveda	1. Ashtaanga hrudaya (अष्टाङ्गहृदयम्) 2. Astanga sangraha of Vaagbhata, 3. Charaka samhita(चरखसंहिता) of Charakhaachaarya, 4. Sushrutha Samhitha(सुश्रुतसंहिता) of Sushrutha and 5. Vrukshaayurveda of Surapala
2	Prosody/Chandas	1. Pingala's Chadashshastra
3	Thesaurus/Nighantu	1. Nirukta (निरुक्तम्) of Yaaskacharya 2. Amarakosha (अमरकोशः) of Amarasimha
4	Upanishads	There are more than 100 Upanishads found but mostly 10 are considered which are 1. Esha (ईश) 2. Kena (केन) 3. Katha (कठ) 4. Munda (मुण्ड) 5. Mandukya (माण्डुक्य) 6. Prashna (प्रश्न) 7. Taittariya (तैत्तरीय) 8. Aithareya (ऐतरेय) 9. Chandogya (छान्दोग्यम्) 10. Bruharaaranyaka (बृहदारण्यकम्)
5	Grammar/ Vyakaranam	1. Ashtadyaayi (अष्टाध्यायी) of Paanini 2. Varthika (वार्तिका) of Katyaayana or Vararuchi 3. Mahabhaasya (महाभाष्यम्) of Patanjali
6	Political economy/ Arthashastram	1. Arthashastra (अर्थशास्त्रम्) of Kautilya 2. Shukraniti 3. Naradaniti 4. Viduraniti
7	Engineering/ Shilpashastram	1. Samaraanganasutradhaaraa (समराङ्गणसूत्रधारा) of Bhoja
8	Architecture/ Vaastushastram	1. Matangavaastu 2. Maya samhitaa 3. Vaasthuratnaakara
9	Law/ Dharmashastram	1. Manusmrithi(मनुस्मृति) of Manu 2. Naradasmriti 3. Parashara Smrithi

(continued)

S N	Subject	Popular texts
10	Chemistry/ Rasashastram	1. Rasahrdayatantra by Govind Bhagwatpad
		2. Srasaratnakara by Siddha Nityanatha
		3. Rasarnava by an unknown author
		4. Srasendracudamani by Somadeva
		5. Rasaratnasamuccaya by Vagbhatta
		6. Rasaprakasasudhakara by Yasodhara
		7. Rasarajalaksmi by Ramesvara Bhatta
		8. Rasendracintamani by Dhundukanatha
		9. Rasendracintamani by Ramacandra Guha
		10. Rasasara by Govind Acarya
		11. Rasakaumudi by Sarvajnacandra
		12. Rasabhesajakalpa by Surya Pandita
		13. Rasasamketakalika by Camunda
		14. Lohapaddhati by Suresvara
		15. Kankaligrantha by Nasirshah
		16. Rasamuktavalina by Devanatha
11	Literature/Sahitya	1. Raghuvamsham(रघुवंशम्) &
		2. Kumarasambhavam(कुमारसंभवम्) of Kalidasa
		3. Kirataarjuniyam(किरातार्जुनीयम्) of Bharavi
		4. Shishupaalavadham(शिशुपालवधम्) of Magha
		5. Naishadheeyacharitham(नैशधीयचरितम्) of Sriharsha
		These are called the 5 'Mahakaavyas' or the greatest among the available ample literary works. The other prominent poets are Bhasa, Bana, Bhavabhuthi, Dandi, Shudraka, Bhatti, Vishaakhadutta, Bhartruhari, etc.
12	Epics/Akhyayika	1. Srimadraamaayanam (श्रीमद्रामायणम्) by Maharshi Valmiki
		2. Srimanmahabharatham (श्रीमन्महाभारतम्) by Bhagawan Vyasa
		3. Astaadashapuraanaani (अष्टादशपुअराणानि) by Bhagawan Vyasa
13	Astronomy/ Jyoutisham	1. Bruhatsamhitha (बृहत्संहिता) of Varaahamihira
		2. Aryabhateeyam
		3. Surya Sidhdhanta
		4. Lilavathi

References

Bhavanishankar BS (2007) Irrigation systems in Vijaynagar empire times and irrigation structures in the old Mysore state

International History Seminar on Irrigation and Drainage, Tehran-Iran May 2-5, 2007, https://www.civilica.com/Paper-IHSID01-IHSID01_002=IRRIGATION-SYSTEMS-IN-VIJAYNAGAR-EMPIRE-TIMES-AND-IRRIGATION-STRUCTURES-IN-THE-OLD-MYSORE-STATE.html

Websites

http://en.wikipedia.org/wiki/Bhagiratha

http://en.wikipedia.org/wiki/Kaveri#Significance_in_Hinduism

http://en.wikipedia.org/wiki/Talk%3AMadikeri

http://www.indialine.com/travel/karnataka/fairs-festivals.html

http://www.nih.ernet.in/rbis/vedic.htm#water_purification

http://www.slideshare.net/ashoknene/environnement-protection-awareness#btnNext

http://www.esamskriti.com/essay-chapters/Traditional-Methods-of-Water-Harvesting-and-applicability-2.aspx

http://www.ssmrae.com/admin/images/0d0c3d83c5f336f7ab602c80838e191f.pdf

http://www.fordham.edu/halsall/india/kautilya1.asp#Book III, Chapter 8

http://en.wikipedia.org/wiki/Sambhar_Salt_Lake

https://www.google.co.in/search?q=images+of+Balsamand+Lake&tbm=isch&tbo=u&source=univ&sa=X&ei=8LWmUrHhJ4exrgeEg4CwCw&ved=0CCoQsAQ&biw=1366&bih=599#q=images+of+Man+sagar+Lake&tbm=isch&undefined=undefined&facrc=_&imgdii=_&imgrc=RmCjLZZWr_bd7M%3A%3BSPwG9Iv98Ps0DM%3Bhttp%253A%252F%252Fwww.bird-fair.org%252Fimages%252F09.jpg%3Bhttp%253A%252F%252Fwww.birdfair.org%252Fman_sagar_lake.htm%3B500%3B228

http://en.wikipedia.org/wiki/Dhebar_Lake

https://www.google.co.in/search?q=irrigation+in+vijayanagara+empire&biw=1366&bih=599&tbm=isch&tbo=u&source=univ&sa=X&ei=1b6mUviEPIGRrQeU94H4CQ&ved=0CD4QsAQ#q=water+management+in+vijayanagara+empire&tbm=isch&undefined=undefined&imgdii=_

http://water.org/solutions/

http://en.wikipedia.org/wiki/Colorado_river_dispute

http://www.salmanmasalman.org/wp-content/uploads/2012/12/InternationalWaterDisputes2.pdf

http://en.wikipedia.org/wiki/Indian_maritime_history

http://en.wikipedia.org/wiki/Harappan_hydraulic_engineering

http://en.wikipedia.org/wiki/Gandaki_River

http://en.wikipedia.org/wiki/Hot_spring

http://en.wikipedia.org/wiki/1993_Latur_earthquake

http://astrouniverse.com/articles/tepotsava.php

http://www.udaipurblog.com/celebrating-the-greenery-hariyali-amavasya.html

http://www.hinduwisdom.info/Nature_Worship.htm

http://en.wikipedia.org/wiki/Kumbh_Mela

http://www.everyculture.com/wc/Germany-to-Jamaica/Rajputs.html#ixzz2ndKvhepY

http://www.iwmi.cgiar.org/Publications/Water_Policy_Briefs/PDF/wpb16.pdf

samskritabh...@yahoogroups.com

http://www.indiawaterportal.org/articles/hydrology-ancient-india-book-national-institute-hydrology-1990

http://www.slideshare.net/D5Z/f3t356
http://ia.rediff.com/news/2005/apr/11brass.htm
http://www.new1.dli.ernet.in/data1/upload/insa/INSA_1/20005b5f_327.pdf
http://www.lonelyplanet.com/india/himachal-pradesh/events/other/minjar-festival#ixzz2qM6RMR2V
http://hinduism.about.com/od/godsgoddesses/a/ganga.htm
http://en.wikipedia.org/wiki/Vallam_Kali#The_most_renowned_of_these_races

The Legal Lore of Water Ecology and Scriptures

Manjeri Subin Sunder Raj

Background

While delving into the finer aspects of history and the growth of various civilisations, the intimate connection between certain aspects undoubtedly crosses the mind. The relation between social and economic progress coupled with the stability of civilisations and people, together with the accessibility and dependability of water resources and water supply, is one such example that comes across many a time, in history. It is based on this aspect that many a historian have deduced the fact and held that the first developed social groupings were hydraulic civilisations.

Whenever recalling the history classes, it would undoubtedly be found that almost all the civilisations would necessarily be attached to and identified with the help of the river basin or river valley around which they developed. This clearly depicts the importance water commanded during that period of time, and it was not by a sheer coincidence that it was termed 'the elixir of life' but was, in all probability, the best way to put forth its importance. Thus, to name a few, the Egyptian civilisation (the Nile civilisation), the Mesopotamian civilisation (the Euphrates-Tigris civilisation), the Harappa and Mohenjo-daro civilisation (the Indus Valley civilisation), the Chinese civilisation (the Huang-Ho civilisation) and so on and so forth.

Thus, it can be seen that waterbodies were considered one of the most important sources of life, while civilisations grew along their banks. Human migrations, the coming into existence of towns and cities, the growth of communities and the advent of agriculture all had imbibed within themselves the importance of water, and it was not by a sheer and mere coincidence that people tended to settle close to well-irrigated and water-rich areas. It is quite fascinating when one realises that water and waterbodies were also a means of transportation which played a significant role in giving

M.S. Sunder Raj (✉)
National Law School of India University (NLSIU), Bengaluru, India
e-mail: subin@nls.ac.in

K.V. Raju, S. Manasi (eds.), *Water and Scriptures*,
DOI 10.1007/978-3-319-50562-6_5

birth to many cities, all around the world, across the banks of rivers. This would inextricably show the relation that water had as regards trade besides fostering relations between various kingdoms.

Methodology

The research work aims to analyse the legal facets of protection, conservation and sustainable management, besides the role, responsibilities, rights and entitlements of all the concerned over water, prevalent in the hoary past. It is also an earnest attempt to examine the extent of their durability and continued relevance in the current context. Keeping this in view, it is pertinent to ensure that a clear idea is given as far as what all would be incorporated in the same. This is done for the very fact that accessing and obtaining authentic materials as far as the topic is concerned has posed quite a challenge. The work aims at bringing out the legal, administrative and regulatory aspects as regards water and water governance, as can be deduced from the other three papers written on the relation between water and religion, namely, Hinduism, Christianity and Islam. While the importance attached to water and the status it enjoys in the religious texts, tenets and teachings have been elaborated on, what has been the aim of this work is to identify those legal and administrative aspects with respect to water management and water governance.

Taking India as a central point, the author has tried to identify the principles that can be applied to the management of natural resources as a whole and water specifically with the cut-off period being 1947.

Although the primary focus of the paper is to establish a common ground, with the other three papers, some independent research also would be attempted so as to bring in some facets of the same that could help understand and analyse the situation more lucidly.

Limitations

While it is known that water was treated as a special resource and that there were quite a lot of regulations, rules and even laws that sought to protect it, many factors tend to cause difficulties in truly fathoming the same. In general, three sources can be mentioned, which have had quite a large influence over the studies undertaken.

1. Interpretations and elucidations made by various historians on the ways of life, conditions that prevailed and the laws that were in place in ancient society
2. Ancient texts, manuscripts and the like which throw light over the various developments in addition to providing insights into the area
3. Legal or other similar documents that have been passed on to the present generation, which necessarily include such principles that were prevalent and were put into practice

Although the advantage of modern-day technology is present, such a study is quite difficult to attempt owing to the fact that such anecdotes, texts and similar other documents that would have helped throw light over the subject have seldom passed the test of time besides being lost. Since the whole text cannot be recovered, in most circumstances, to join the dots and come up with a comprehensive picture would be practically impossible. It is also the case in most of the situations that, even if the whole text could unearthed, the language used would be hard to decipher and even if deciphered with utmost difficulty, might not be able to convey the actual meaning and purpose for which it was enacted. The lack of knowledge of the situations prevalent during that period of time is also a major hindering factor to the real understanding of the then prevailing law. To have understood that law, one should have been able to comprehend a myriad dimensions, which unfortunately would not be found in a single person and even if found would necessarily suffer from the lack of adequate and reliable data.

Need and Necessity

The importance that such primitive and age-old traditions hold and the guidance that they provide cannot be overlooked and for that very fact, studies have been initiated to unearth, comprehend and imbibe those traditions and laws that we had in place way back and ways to explore the very essence of them for general good.

This study intends to explore the various laws and legal systems prevalent in the past which had a direct link to water and water resources besides acting as a precursor to the water rights that are in vogue presently.

Conceptualisation of Water: The Life-Giver

To live in harmony with nature has been one of the cardinal principles that have guided man from time immemorial. Although it can be said that man was afraid of nature and natural phenomena and treated them as somewhat of a superior power, it was only when he stopped his nomadic ways of life that a true comprehension of nature's power and importance came to be conceived by man. While trying to be in tune with nature and to harness her various gifts, man came to understand that Mother Nature was the sole reason for his survival.

Such an affinity that was present with nature helped man understand the various gifts of nature and put them to judicious use. While delving deep into the Indian scriptures, the Pancha Bhoota concept[1] helps one realise the sacredness that was

[1] This concept puts forth the idea that matter is made of five elements, namely, earth (*Prithvi*), water (*Jal*), fire (*Agni*), air (*Vayu*) and ether (*Akasha*). For more, see http://www.indian scriptures.com/vedic-lifestyle/beginners-guide/pancha-bhootas-or-the-five-elements, accessed on 03/06/2016.

assigned to nature and the reverence that was attached to each and every part of it. To protect it was considered the sacred duty of all, for she was the provider, without which we would not be alive.

Respecting nature and according her all powers and living in harmony with her had been conceived way back in the past. While the right to use her produce and gifts was present, the important aspect was that such gifts were to be used only to satisfy one's needs, not anything more. Destruction of nature was seen akin to the destruction of mankind itself and for that fact, it was quite important that she be treated with utmost care, reverence and respect. Attributing divine qualities to Mother Nature was a way in which such reverence, care and respect were sought to be achieved and that played a great role in the protection of natural resources.

Various natural phenomena, like earthquakes, lightning, floods, storms, droughts, volcanic activity, torrential rain, etc., were treated as manifestations of anger and wrath by Mother Nature, and thus man was afraid of it from time immemorial. It was only when the science behind these occurrences was understood that he came to realise and understand the actual physics behind it. Even then it was evident that such actions of man as would tend to be harmful to Mother Nature would have played a role in such occurrences. Thus, somewhere down the line, the idea of conservation and protecting the resources for well-being grew in stature.

Water: The Resource

'Life-giver and life-sustainer' are common expressions associated with water. Quite true as it is one of the most important resources for the survival of mankind.

Having been allocated an important position in nature, it was quite natural that water was widely accepted as the most important of all the resources that mankind was part of. Treating it as a divine gift, it was common for water to be accorded such high status, so as to ensure that it was equitably distributed and taken care of. As was discussed earlier, the very fact that it being the elixir of life made it so important that protecting and preserving it was considered highly important. The status that was accorded to water was, therefore, quite high.

It is to be understood from this very fact that quite a lot of thought was accorded to how this valuable resource was to be put to use and the measures to be adopted so as to ensure its availability, equity and accessibility, throughout the ages. Highly circumspect was its use, due to the importance that it had.

One of the most striking factors of water conservation is that there was an element of divinity; rather the religious character of water was given prominence. Such an element ensured that a resource as important as water would not be subjected to misuse. This would have definitely helped a long way in ensuring that the resource was not put to bad use and that people treated it as a gift of God and thereby revered it and tried not to put it to bad use. This would have undoubtedly ensured that water as a resource was well protected and used judiciously.

Although it can be seen that there was an element of divinity as attached to water, it is also evident that the King, being treated as an embodiment of the divine being, had the power to make laws that set forth the guidelines for protecting the resources. Thus it can be understood that certain safeguards were envisioned which were supposed to have the effect of augmenting and protecting water and water resources and also preventing them from being polluted. This was also envisaged in a much larger canvas wherein steps were also taken to store the resource and put it to an equitable use and distribute it among the people in an equitable manner so that all had their share of it. Conservation of the resource for future use and future generations too was present in its **rudimentary** form, for that was the very fact why it was accorded such a status.

While delving into history and accounts given by various historians, such a situation can be understood wherein there was quite an imminent need on the part of the people, wherein a control over water usage was essential. This was to ensure that the resource was put to use in such a way that the demands of all the people were satisfied with an equitable distribution of the same being in place. It is also interesting to note that such an equitable distribution of the same was to be done not only in the cases of various users that were present but also as regards various uses to which it was utilised. Many a time, this was the sole criterion for which there arose conditions and situations which guaranteed the protection of water.

The very reason behind this was the very fact that this resource was treated as a common resource; a resource over which everyone had a right. Water was treated as a source of right; a source of entitlement that was prevalent and present in all. However, there were also situations wherein certain people were kept out of accessing water. This was to ensure that the resource was well maintained and that it was not put to waste.

Since the concept of rights attached to the use of water was prevalent, there arose the need to identify the duties too to be passed on and be prevalent among not only the state but also other stakeholders. Water, treated as a common right of all, thus assumed significance, hitherto absent.

What we have to understand from this fact is that it is from such a situation that the need for the world's first laws on water, water systems and water governance arose. Based on the nature of their location, i.e. their physical, geological and climatic conditions that prevailed, the social and economic conditions that prevailed during those times and areas, and the philosophical and religious tenets and beliefs of the people concerned, such water laws grew in stature, importance and character.

Based on the availability of water and its abundance, along with its periodic variations, different aspects of water control were exercised. This would indubitably throw light on the importance attached to water and the role that such a resource was to play in the growth of human civilisation. Had it not been for water, civilisations would never have flourished this far.

Based on the assertions and points identified, it was sure that the growth and development that such old civilisations were able to reach at would necessarily be related to the efficacy and efficiency of various controls—organisational, spiritual and legal—that were prevalent at that point of time. This argument strengthens further on realising that many a civilisation that did not attach much importance to these aspects disappeared without a trace over a period of time, unable to withstand the rigours that were put across due to water scarcity and similar problems.

The importance attached to water, as far as civilisations were concerned, grew further due to the fact that during times of invasions and attacks, waterbodies, water channels and water works were among the first to be targeted, so as to mount pressure on the enemy. It is quite familiar to us that in history, wars would have necessarily be won owing to the victorious side taking advantage of the situation by destroying key structures thereby diminishing the resources. Water, naturally, was quite high up that list and was always sought to be protected. This was the sole reason that even if the city was well fortified and the people were safe behind the walls of the city, there arose a need to have a dependable and continuous water supply system to the city, which could not be disrupted by the enemy. This explains as to why much thought and engineering skills were put to use while creating water channels and water systems, which surely were the lifeline of the city and thereby the civilisation.

Based on all the above assumptions and assertions, what can be deduced is that the religious scriptures that felt the pulse, rather, which conveyed the pulse of the society, ensured and tried to imbibe the ways in which augmentation, protection and prevention of pollution, storage, distribution and conservation of water and water resources could be enforced voluntarily.

Making the people themselves aware of the importance of such resources and thereby inculcating in them the very spirit of protecting the same was the method that was to be tried and tested and hopefully pass the test of time.

Delving into the right aspect of water, the fact that it still exists as part of the customary laws, clearly shows that the same are not purely legal rights but are those rights which have been recognised in law.[2]

Legally, Water: An Articulation

As has been pointed out, earlier, political, geographical as well as sociological influences played a great role in shaping the laws, doctrines and legal articulation related to water. With the growth in importance as far as water was concerned, it has to be understood that quite a lot of changes took place, necessarily bringing into the picture various doctrines and theories that can be understood to have facilitated 'water jurisprudence'.

The theories so formulated can be divided into two, based on the period in which they had been formulated, i.e. traditional doctrines and modern doctrines. For a more clear picture, the traditional doctrines that were present and the evolution of the same which led to the modern doctrines of jurisprudence need be looked into in detail.

---

[2]Chatrapathi Singh, *Water Rights in India*, in Chatrapathi Singh (Ed.), Water Law in India, ILI, 1992, at p. 11.

Traditional Doctrines of Water Jurisprudence

Coming to the growth of water jurisprudence in olden times, there were three main doctrines prevalent—the Doctrine of riparian rights, the 'first come, first served' principle or the prior appropriation theory and the territorial sovereignty theory or the 'servitude' theory.

The Doctrine of Riparian Rights

To put it in other words, the *Finders keepers, losers weepers* principle is one of the oldest principles related to the use of water. According to this principle, only those whose property had water flowing through it or flowing adjacent to it had the right to use that water.[3] This doctrine was part of Roman law. It was applied in times when there was plentiful water. However, over time, the same became redundant due to the fact that transportation of water and altering the course of water systems became easier.

Another defect which this doctrine had can be understood by going through Garrett Hardin's article, *Tragedy of the Commons.*[4] If all the users tried to utilise the resource to its maximum, then there would be a reduction in the resource as such, leading to a situation ultimately wherein the resource itself would be lost.

Distributive justice principle, i.e. in this case, a resource that is naturally available which should be used by all, is also not followed according to this doctrine. Just because there is a waterbody which is close to one's property does not necessarily explain as to why that person should be allowed to have a superior right over those holding no property nearby. The doctrine has been excluded in cases where there have been serious conflicts with respect to the distribution of water, due to this very aspect.[5]

As far as Indian law is concerned, this doctrine has been discarded by all the water tribunals. The Krishna Water Disputes Tribunal has observed that the doctrine, as is present in the Indian Easements Act, 1882,[6] could be applied to individuals but is not useful while considering interstate water disputes.[7] The above views were confirmed by the Godavari Water Disputes Panel in 1979.[8] The Eradi (Ravi and Beas) Waters Tribunal in 1987 also rejected the doctrine by putting forward the idea and showing as to how the doctrine failed to support the principles of justice.[9] The Narmada Water Disputes Tribunal of 1974 also rejected the doctrine.[10]

---

[3] Santosh Kumar Garg, International and Interstate River Water Disputes & Interlinking of Rivers, Khanna Publishers, New Delhi (2003).

[4] Garrett Hardin, *The Tragedy of the Commons*, 162 Science 1243 (1968).

[5] *Supra* n. 3.

[6] Sec. 7(b), The Indian Easements Act, 1882.

[7] The Report of the Krishna Water Disputes Tribunal, Vol. 1, 93 (1973).

[8] Tony George Puthucherril, *Riparianism in Indian Water Jurisprudence*, in Ramaswamy R Iyer (Ed.), Water and the Laws in India, Sage Law, New Delhi, 2009, at p. 97.

[9] *Ibid.*

[10] Ramaswamy R Iyer (Ed.), Water and the Laws in India, Sage Law, New Delhi, 2009.

The Doctrine of Prior Appropriation

Based on this theory, all water is considered as a public property while it is in its natural course. This further puts in that the lawful right to use it may be attained by appropriation and application for beneficial use. This theory is also known as the '*first come, first serve*' theory, and it states that such 'prior' water rights last forever, and changes in the situation would not be reflected.

The doctrine too has quite a lot of inherent flaws that makes it not possible to solve disputes relating to water. It can also be termed as arbitrary in nature due to the very fact that the political feat of conquering a certain territory is taken into consideration over and above the principles of distributive justice.

The resource is treated as a private property and not as a community resource by this doctrine, which can be said to be one of the most derogative aspects. It can be seen that due to such a practice, the doctrine favours a private property regime. There is some sort of a private control over the right to use as well as the right to alienate the resource. These have rendered the same redundancy in present times.

This doctrine was not put to use in India and the Calcutta High Court in *Belbhadra Prasad* v. *Sheikh Barkat*[11] rejected the same. Even though in the USA, the case of *Wyoming* v. *Colorado*[12] used this doctrine, it was at a later point of time rejected in *Connecticut* v. *Massachusetts*.[13]

The Theory of Territorial Sovereignty

This theory states that a riparian state does not have a right to demand a continued flow of water from another state. This theory, which is also called the Harmon doctrine, is based on two principles:

1. The might is right principle
2. The 'slavery' or servitude principle

The first principle or the doctrine of absolute territorial sovereignty owes its origin to Judson Harmon.[14] He, in 1896, declared that the USA would not share the waters of the Rio Grande River with Mexico. This declaration was made on the basis of the fact that water flowing in the territory of the USA belonged to the USA.

The servitude principle has its origin in the Roman law. The position was that whatever was attached to one's property was to serve the owner of the property. The owner was free to do whatever he pleased as he had the absolute right over them.

[11] 11, C.W.N. (1906-7) 85.

[12] 259 U.S. 419 (1922).

[13] 282 US 660 (1930).

[14] Mr. Harmon, a democratic politician from Ohio was appointed as the US Attorney General under President Grover Cleveland, on June 8, 1895. He also served as the 45th Governor of Ohio.

Due to the very nature of this theory, in modern days, it is not used. However, it is quite interesting to note that as regards the Columbia River, wherein the USA was the lower riparian as opposed to Canada, the doctrine was rejected by the USA itself, thereby recognising the rights of the lower riparian. The Harmon Doctrine was also rejected by the international community later on.[15]

This doctrine is not accepted in India. If we look into the Indus Waters Treaty of 1960, it becomes obvious that Pakistan was entitled to use the waters of Indus, Jhelum and Chenab, the rivers in the west, while India could use the waters of Ravi, Beas and Sutlej, the rivers in the east. This was done so in spite of the fact that India was the upper riparian with respect to all the rivers. Had the doctrine been used, Pakistan would not have been able to use the waters of any of the rivers.

Evolution of Doctrines of Water Jurisprudence

Since this paper tries to examine the principles of the past, it is to be kept in view that the principles of water jurisprudence, dealt with earlier, have undergone a sea change, giving way to newer doctrines that are prevalent now. Although the scope of the present paper does not include the present principles, a brief description of the same, it is felt, is inevitable; so as to ensure that it is known as to how the traditional principles that were in practice in the past slowly gave way to newer principles. The elements that were present in earlier doctrines evolved and took a new shape based on the needs and necessities and with the passage of time emerged in newer forms, albeit better. A brief description of the same is attempted, so as to gain some insights into the paradigm shift that took place. Whether the old principles gave way to new principles in one go or whether there was a continuous evolution or a modification of the old ones which gave rise to the new ones need be looked into.

The equitable apportionment theory, which is based on the ideas of equality and fairness, can be said to treat all claimants as equal right holders with apportioning the resource as per their individual needs through fair and legal means. The utilitarian concept, proposed by Jeremy Bentham, which puts forward the 'greatest good for the greatest number' can be said to be the guiding principle of this theory.[16]

An important feature of this concept/theory is that the doctrine itself provides a choice to decide on the basis of needs or merits. However, it is quite interesting to note that in almost all the cases, the authorities having powers to decide preferred the basis of the needs of the parties involved rather than merits. To consider an example related to India itself, it can be seen that, prior to independence, the Indus River water was apportioned on the basis of equitable needs of both Punjab and Sindh. Further, in independent India too, the same has been followed.[17]

---

[15] The Report of the 52nd Conference of the International Law Association, Helsinki, Finland, 1966.

[16] S. Basheer Hussain, *The Law of Inter-State Rivers in India: Principles of Equitable Apportionment of River Waters*, 17 IJIL 43 (1977).

[17] *Supra* n. 10.

The Helsinki Rules of 1966 laid down various factors to be taken into consideration in the beneficial uses of the waters of an international drainage basin.[18] In USA, a number of judicial decisions, for example, like the 1930 judgement in *Connecticut* v. *Massachusetts*,[19] overruled the prior appropriation doctrine and favoured the settlement of disputes on the basis of the principle of equity. Down the years, in the cases of *New Jersey* v. *New York*,[20] *Nebraska* v. *Wyoming*[21] and *Arizona* v. *California*,[22] this principle was strengthened.

The distinction between the two terms needs be looked into at this juncture: 'equitable apportionment' and 'equal apportionment'. Under the latter, water can be used equally by both the upper and lower riparian states. But under equitable apportionment, many relevant factors would have to be taken into account. It is also interesting to note that in such cases, a neighbouring non-riparian may also have a right to use the water to a very limited extent.[23]

Although this can be said to be superior to the riparian rights theory, it has its own set of failings. A major point of worry is to find an authority that is acceptable to all parties, especially in respect of international water disputes. However, in the case of national disputes, this is not the case as there would, generally, be an adjudicating authority created by the legal systems within the territory.[24]

---

[18] Article V of The Helsinki Rules on the Uses of the Waters of International Rivers, 1966, available at http://www.unece.org/fileadmin/DAM/env/water/meetings/legal_board/2010/annexes_groundwater_paper/Annex_II_Helsinki_Rules_ILA.pdf, last accessed on 03/06/2016.

I. What is a reasonable and equitable share within the meaning of Article IV needs to be determined in the light of all the relevant factors in each particular case.

II. Relevant factors which are to be considered include, but are not limited to:

1. The geography of the basin, including, in particular, the extent of the drainage area in the territory of each basin state
2. The hydrology of the basin, including, in particular, the contribution of water by each basin state
3. The climate affecting the basin
4. The past utilisation of the waters of the basin, including, in particular, existing utilisation
5. The economic and social needs of each basin state
6. The population dependent on the waters of the basin in each basin state
7. The comparative costs of alternative means of satisfying the economic and social needs of each basin state
8. The availability of other resources
9. The avoidance of unnecessary waste in the utilisation of waters of the basin
10. The practicability of compensation to one or more of the co-basin states as a means of adjusting conflicts among uses
11. The degree to which the needs of a basin state may be satisfied, without causing substantial damages to a co-basin state

[19] 282 US 660 (1930).

[20] 283 US 336 (1931).

[21] 325 US 89 (1945).

[22] 373 US 541 (1963).

[23] *Supra* n. 16.

[24] *Ibid*, at p. 45.

Further, the equitable utilisation theory, based on the Helsinki Rules, 1966,[25] in comparison to the equitable apportionment one, which has a third party adjudicator, gives the two parties enough flexibility to choose their own way of dispute resolution. Moreover, it can also be seen that this theory is more concerned with the resource itself, whereas the equitable apportionment theory is based on the needs of the people as well as the state. Due to this very fact, this theory favours maximum as well as optimum utilisation of the same and, thereby, can be opined to follow the Benthamite concept of utilitarianism. However, this too suffers from the same problems as those faced by the equitable apportionment theory.

The community of interests theory, a modified version of the equitable utilisation theory wherein the term community means concerned claimant states, is also one that has evolved in modern times. Earlier, there was a difference between *bonum vacans*[26] and *public juris*,[27] which was discussed in *Embrey* v. *Owens*[28] and accepted as the traditional Roman law distinction. The former was termed as 'negative community', whereby the new communities taking interest in the common property resources could substitute it by new communities of shared interests. As this theory allows the groups participating in the distribution to be defined as communities, the same has an advantage over the equitable utilisation theory. However, this too suffers from the same problems as those of equitable utilisation.

All these principles have played a great role in shaping the way in which water has been put into use. The modern principles have necessarily paved the way for promoting the augmentation and sustainability of water as well as water resources. By way of putting into practice these principles, various legal systems have ensured that the resource is well maintained and taken care of for future generations.

Principles of Governance

Since, like air and food, water too being an essential aspect of human survival, distribution and acquisition of the same has been subject to legal concerns, from olden times. The main difference that can be noted is that in ancient times, when the same was available in vast quantities, there was not much of a need to control the use of the same. This was the main reason as to why the 'principle of discovery'[29] was in place. Since during those times, the resources were aplenty and there was never a burgeoning demand, this did not lead to a crisis. However, with the passage

[25] Article IV and V of The Helsinki Rules on the Uses of the Waters of International Rivers, 1966, available at http://www.unece.org/fileadmin/DAM/env/water/meetings/legal_board/2010/annexes_groundwater_paper/Annex_II_Helsinki_Rules_ILA.pdf, last accessed on 03/06/2016.

[26] Nobody's property.

[27] Everyone's property.

[28] (1851), 6 Exc.35.

[29] By virtue of this, anyone who was the first to 'discover' was supposedly the owner and enjoyed dominion over the same.

of time, when there was a scarcity of resources, thanks to over-exploitation and uncontrolled use, there arose an urgent need to bring about some sort of an external control.

While talking of governance, it can be understood as an *exercise of economic, political and administrative authority to manage a country's affairs at all levels; it comprises the mechanisms, processes and institutions through which citizens and groups articulate their interests, exercise their legal rights, meet their obligations and mediate their differences.*[30]

Environmental governance is a concept in political ecology and environmental policy. Sustainability as the supreme consideration for managing all human activities—political, social and economic—is advocated by the same.[31]

While delving into all these concepts, it is quite imperative that one looks into the basic one, the one which conveys the idea of the rest, i.e. natural resource management. This has been the talk of the town for quite a long time, and it has taken an adequate care and caution so as to ensure that such resources, of which no individual is the sole owner, is put to use in the most judicious and equitable manner so as to ensure its availability, accessibility and management.

Taking into consideration the human rights angle, it was only a matter of time before it was recognised that the management of natural resources should be in tune with the resource state as such besides being able to ensure the principle of equity as far as use and management was concerned.

In this chapter wherein the resource, water, is discussed, the governance of it can be understood to be *the range of political, social, economic and administrative systems that are in place to develop and manage water resources and the delivery of water services, at different levels of society.*[32]

As it signifies, *water governance* carries quite a lot of meaning in terms of portraying the value attached to it. As discussed earlier, the ramifications it has on quite a lot of arenas, be it political, social, economic or even administrative, has been subject to quite a lot of scrutiny.

While taking into consideration the principles that have governed the use of water, it is quite natural to come across different ones based on geography, history of the place, social and other cultural aspects and the like. However, it is also interesting to note that there are threads of unison present so as to show us that there has been some sort of an overlapping of such principles of governance. This has been an outcome, a direct one too, of the various inclinations attached to water.

Historically too, one finds a lot of examples of factors that have led to the convergence of such inclinations. Whether it is the growth of various civilisations at differ-

[30] United Nations Development Programme 2001, available at http://www.un.org/millenniumgoals/pdf/Think%20Pieces/7_governance.pdf, last accessed on 03/06/2016.

[31] Oliver M Brandes and David B Brooks, The Soft Path for Water in a Nutshell, Friends of Earth Canada and POLIS Project on Environmental Governance, University of Victoria (2005), at p. 5.

[32] Global Water Partnership 2002. See also Peter Rogers, Alan Hall, Effective Water Governance; TEC Background Papers No. 7; Global Water Partnership: Stockholm, Sweden, 2003, available at https://dlc.dlib.indiana.edu/dlc/handle/10535/4995, last accessed on 03/06/2016.

ent points of time or the spread of such civilisations over a large area, such forces have helped a lot in bringing forth various principles of governance.

It is interesting to note that apart from the rise of such civilisations, the spread of them, too, coupled with factors such as trade, colonisation, conquests and religious factors played a great role in shaping such principles.

With the passage of time, there emerged a lot of principles that actually guided and served as a beacon for enhancement of such rights. According to Prof. Philippe Sands,[33] the main principles that can be identified as far as environmental governance is concerned are 'polluter pays' principle,[34] 'precautionary' principle,[35] principle of 'sustainable development',[36] 'public trust' doctrine[37] and the principle of intergenerational and intragenerational equity.[38] That apart, there were also a few other concepts, the basis of which can be found in history. While the whole of earth and resources were treated to be for the living beings as such, the idea of commons and common property resources took shape. It can also be found in the religious texts with references made to man being cast upon a higher pedestal of duty due to his superior position and the concept of him being treated as a steward, i.e. a conservator of resources bestowed on living beings. A brief discussion of the various concepts and principles is hoped to help illuminate the context in which they were put to use.

Commons

The idea of commons or common property resources need be delved into as they are significant. Since time immemorial, when resources were available to all, it was a matter of who would use the same. It can be seen that ways of appropriating a resource were prevalent. However, there were resources that were treated as part of the wider environment, and it was a practice in vogue that all had a right to use such resources. This is when the idea of commons came into the picture. Developed by Hugo Grotius, this idea soon cottoned on.

While considering the major impact that this idea had on the community, it can be seen that the most important was the one in which the said resource was treated as commonly owned. No one in particular or any community had an overriding and absolute control over such resources. It was treated as a resource over which the rights were equally divided, and there also was a duty to ensure that the resource was protected. While water definitely would come under the scope of the same, the importance attached to it depicts the level of protection that was accorded to it. This idea that tried to portray the resource as being provided for the entire gamut of life

[33] Philippe Sands, Principles of International Environmental Law, Vol. 1 183 (1995).

[34] *See M.C. Mehta* v. *Kamal Nath*, (2000) 6 SCC 213.

[35] *See Vellore Citizens' Welfare Forum* v. *Union of India*, (1996) 5 SCC 647.

[36] *See Goa Foundation* v. *Diksha Holdings Pvt. Ltd*, (2001) 2 SCC 97.

[37] *T.N. GodavarmanThirumulpad* v. *Union of India*, AIR 2003 SC 724.

[38] *State of Himachal Pradesh* v. *Ganesh Wood Products*, (1995) 6 SCC 363.

on earth did exert a considerable influence over the use of water. As is understood and conveyed in various religious scriptures, water was treated as an elixir of life, and, hence, was accorded the utmost importance and reverence.

Stewardship

Man has been cast upon with an immense duty. Having been created as a superior being, it was only a matter of time before realising that with a greater power accorded to him, a greater responsibility too he had upon his shoulders. It is this responsibility that made him take steps and come up with actions to protect the environment. Stewardship is a careful and responsible management of something (in this case water) entrusted to one's care, it can be said.

The concept of Bhoopala, as discussed in the Hindu scriptures, stems from this very concept of man being treated as a 'palaka', i.e. a servant or a trustee. Literally it conveys the idea that man takes care or is supposed to take care of the environment and everything that it encompasses. This consists of a rudimentary structure wherein man is entrusted the responsibility of environmental protection.

The Bible too makes clear references to the duty cast upon humans wherein they are caretakers of God's creations. As is made clear from the paper on Christianity, environmental stewardship and care for all the creatures of the earth is commanded throughout Genesis. The dominion that has been granted to mankind in a way increases the responsibility that he holds.

As is evident from the paper on the Islamic perspective on water, Allah created humans so that they can act as the guardians of His creation, which, in turn, puts forth the same idea of stewardship wherein we humans are actually only trustees. This trusteeship or *khalifa* ensures that we cannot do anything that we wish with nature. It is entrusted by Allah for our safekeeping. Upon this title of 'steward' being conferred upon man, it is affirmed that man executes Allah's commands and is a manager of the earth and not a proprietor; he is a beneficiary, and not a disposer. Thereby man can put to use earth and its resources in accordance with the purposes for which it was created and can fulfil his interests but need to ensure that there exists a balance and a control over such use.

Being a steward, a modest moral code is cast upon man, and he is to ensure that the resources are not put to waste. He is also to ensure that the resources are made available to the next generation as well.

Polluter Pays Principle

This principle conveys the idea that polluting the environment entails a duty on the part of the polluter; a duty that he has to perform. Based on the idea that has been conveyed by this principle, the polluter is the sole person responsible for pollution and damage caused to the environment, and it is only a poetic justice that he be made responsible for restoring the environment to its earlier condition.

While achieving pristine conditions might not be feasible or might turn out to be highly impossible, it is highly necessary that the cost of bringing back the environment to the condition that it was in before being polluted be borne by the polluter. He necessarily has to pay for the damages caused and also find methods and ways that can be used for reducing the damage, thereby restoring the same.

Having been adopted by various courts, this principle has been tried and tested and has paved a revolutionary path towards ensuring justice to the environment in terms of providing for money and methods to bring it back to normal.

Precautionary Principle

'Prevention is better than cure' is an adage that we all are familiar with. The same forms the basis of this principle, too, which states that those actions that might cause harm to the public or the environment should be refrained from. If such actions are taken, the person involving in such actions would be held liable.

Anticipation of harm before it is caused, a caution that is exercised, forms the basis of the same. Having been adopted worldwide under several environment-related conventions, this principle surely does throw light on the fact that a necessary precaution is necessary before taking up any step.

Public Trust Doctrine

This is an ancient doctrine, which took birth in ancient Roman law, governing the management of natural resources. It was also part of the law in mediaeval England as well as the USA later.

Unlike the common perception that the state is the 'owner' of water, this doctrine conveys the idea that the state is a trustee of all natural resources. This is to ensure an equitable distribution as per the present needs of various groups, while taking care of the needs of future generations at the same time. The state executes its duties by legislating on, regulating and managing water resources.[39] It imposes limits on governmental actions while providing public access rights to resources, coming under the trusteeship.

In the USA, this doctrine has three important attributes. Due to the same, for the management and distribution of natural resources, this doctrine is put to a wide practice. Firstly, it asserts legal rights among the general public without owning

[39] Joseph Sax, *The Public Trust Doctrine in Natural Resource Law: Effective Judicial Intervention*, 68 The Michigan Law Review, 471 (1970). An online version is available at http://www.uvm.edu/~gflomenh/ PA395-CMN-ASSTS/articles/sax.pdf, accessed on 20/02/2014.

specific property rights. Secondly, it enables action that is enforceable against the government. Thirdly, it enables interpretation consistent with evolving concerns regarding the human environment.[40]

Keeping these in view the government assures a sustainable use of resources. One of the landmark cases in which this doctrine has been put to use is *The National Audubon Society* v. *The Superior Court*, also known as the *Mono Lake Case*.[41] The facts are that due to an uncontrolled and consistent drawing of water, the water level of the Mono Lake in LA decreased. This also had an effect on the lake ecosystem as the water became more saline. The California Supreme Court, invoking the public trust doctrine, brought about a stop to this. It is to be noted that such a circumstance and a ruling were certainly an exception to the general trend, which preferred to be favourable to economic considerations over environmental ones.[42]

While coming over to the Indian system, it can be seen that the ancient concept of 'Bhoopala' had the rudimentary aspects which formed the basis of the above-mentioned principles. The fact that there is a need and necessity to take care of the resources leads us to a situation that we cared for the resources. The very concept of Bhoopala can be said to have been born from this. What Mother Earth provides us with, we are supposed to take care of the same and ensure that it is not lost.

The role of the state as a regulator of such activities indulged in by man also gets shaped from this very concept. The state, it is understood, is obliged to take steps so as to ensure that the resources are well maintained and protected and also that it is available to all.

The very state has to ensure that 'right to water', a basic human right, is available to all. The principle of equity was treated as the basic principle of water governance in the past. It is based on this principle that there was a distribution of the resource in a discrimination-free manner. It can be seen that with the application of this principle in India, there arose the much needed equitable distribution of water among the people. In ancient India, which was characterised by various shades of discrimination based on caste and religion, it was the applicability of this principle that made it possible for everyone to have an equitable and easy access to water.

It is the growth of such principles that helped an equitable and appropriate distribution of water as a resource. Had it not been for the same, the world would not have had the resource for long.

---

[40] Carole Necole Brown, *Drinking from a Deep Well: The Public Trust Doctrine and Western Water Law*, 34 Fla. St. U. L. Rev. 1 (2006). An online version of the same is available at http://www.law.fsu.edu/journals/lawreview/downloads/341/brown.pdf, accessed on 20/02/2014.

[41] 33 Cal 3d 419.

[42] T N Narsimhan, *Water Law for India: Science and Policy Perspectives, 500-551* in Ramaswamy Iyer (Ed.), Water and the Laws in India, Sage Law, New Delhi, 2009.

Sustainable Development

The Brundtland Report—Our Common Future: Report of the World Commission on Environment and Development, Chapter 2, Towards Sustainable Development—puts across that 'Sustainable development is development that meets the needs of the present without compromising the ability of future generations to meet their own needs'. It contains within it two key concepts:

- The concept of needs, in particular, the essential needs of the world's poor, to which an overriding priority should be given
- The idea of limitations imposed by the state of technology and social organisation on the environment's ability to meet the present and future needs

It can be understood that this principle acts as some sort of a beacon light in terms of guiding the mankind towards ensuring that the resources are put to a prudent use and protected for future generations. It strengthens the idea that whatever earth provides us with is not the sole property of one single generation, but it is for the benefit of the whole of the human race that is yet to roam. Each generation is cast with a duty that they are to pass on the resources to the next, thereby ensuring the continuation of life in all forms on this earth.

Intragenerational Equity and Intergenerational Equity

Based on the various principles discussed earlier, the concept of inter and intragenerational equity assumes a greater importance. Through the concept of sustainable development discussed earlier, this concept's crux can be understood. Equity being the core, this principle brings out the importance of natural resources including water that man shares not only with the other living beings living alongside him but also those from the future generations.

Such is the relevance of this concept that man, at every turn, understands as to how important the resources are and how short supplied they are. Based on such facts, man should be able to work his way out of the precarious position that he finds himself in at present and ensure that all is not lost.

Legal Character of Water

Taking into consideration the fact that India had both individual as well as group rights over water, it can be understood that the Indian society had a clear cut management framework for the same. On who the 'water rights' should be vested with is a central question that needs to be tackled in this age of water scarcity.

While there may not be a unison of ideas and a meeting of minds regarding the same, it is quite imperative that we understand that water rights being legal rights have been managed effectively since ancient times. While it can be said that the most prominent among the water laws we had were those of the Romans, modern notions of water rights, it can be said, were influenced by the European law. The common tradition prevalent in England was applicable as far as India was concerned. However, it can be found that even in ancient times, India law was prevalent as far as the management of water was concerned.

Ancient Indian Law

One has to understand the basic tenets and concepts that are prevalent in Hinduism to understand the concept of environment and environmental protection in ancient India. The Hindus believe in the concept of *Brahman*.[43] It is believed that all gods, goddesses and everything else that we see in the universe are a manifestation of *Brahman*. It covers from subatomic particles to mightiest system of stars, planets and galaxies. It has various types of matter, radiation and *gunas*[44] with the characteristics of purity, activity and passivity. These *gunas* are subtle in their nature. It is to be noted that the universe contains both animate as well as inanimate life forms. It is due to this fact that Hindus see divinity in everything.[45] So they believe that all these divine forces have played an important role in sustaining life on this earth. In order to please God, so that the same relation continues, the Hindus believe that they are to live in harmony with His creation. It can be seen that these thoughts have given rise to various customs and practices that are still followed by traditional Hindus.[46]

The earth[47] is considered to be a manifestation of divinity, and, hence, she must be treated with respect. There are a lot of principles that are present in Hinduism

[43] This means an 'all-pervading divine force'.

[44] It is a primordial matter. It means 'string' or 'a single thread or strand of a cord or twine'. The three primary gunas are generally accepted to be associated with creation (*sattva*), preservation (*rajas*) and destruction (*tamas*) http://en.wikipedia.org/wiki/Guna, accessed on 16/10/2013.

[45] It is clear from the Vedas that nature worship was prevalent in ancient India. For an in-depth analysis, see Dr. Raja Ram Mohan Roy, *Vedic Physics, Scientific Origin of Hinduism*, Golden Egg Publishing, Toronto, 1999; Yudhishthira Mimansaka, *Vaidika Siddhanta Mimansa*, Sonipata, 1976; A.R. Panchamukhi, *Socio-economic Ideas in Ancient Indian Literature*, Rashtriya Sanskrit Sansthan, New Delhi, 1998; Ranchor Prime, *Hinduism and Ecology: Seeds of Truth,* Cassel Plc, London, 1992; Dr. Sashi Tiwari, 'Origin of Environmental Science From Vedas', available at http://sanskrit.jnu.ac.in/conf/stait/uploaded_papers/Shashi-Tiwari.pdf, accessed on 08/04/2013.

[46] It can be seen that before the foundation of a building is dug, a priest is invited to perform *Bhoomi Pooja*. This is to worship and appease Mother Earth and seek forgiveness for violating her. For more, see http://www.boloji.com/index.cfm?md=Mobile&sd=Articles&ArticleID=3634, last accessed on 03/06/2016.

[47] Rig Veda 3-55-14 describes nature as Mother Prakriti, and Yajur Veda 18-30 describes Earth as an adorable immortal mother. Earth in the Vedas is 'Prithvi Devi', and it is widespread, very kind and gracious mother.

that play a great role in governing the relation between the people and the environment. Certain cardinal principles are expected to be followed by all Hindus.[48]

- The five elements, namely, space, air, fire, water and earth are the foundation of an interconnected web of life.
- Dharma often translated as 'duty' can be extended to include our responsibility to care for the earth.
- Simple living is a model for the development of sustainable economies.
- Our treatment of nature directly affects our karma.

Protection of Water

The importance that water holds has been clearly understood,[49] and every step has been taken so as to ensure that it is kept pure and free from overuse and exploitation. The life-giving ability of water is recognised in Rig Veda.[50] What has to be kept in view is that recognition is given to the life-providing capacity of water, not only physically but also spiritually. Water being one of the most basic needs is quite natural that there are quite a lot of references made to the same. It is said to represent splendour[51] and bear off all defilements and cleanse people.[52] It is said that water is to be freed from defilement.[53]

The importance of water as a cleansing agent can be understood from the fact, that is, water cleanses humanity of the evil of pollution committed by it.[54] The fact that water should be free from pollution is made clear by the verse that '*Water and herbs should have no poison*'.[55] Water, as it occupies an important role in the life of humans can be said to be one of the gifts without which we would never have survived. The verse '*offerings are dedicated to waters of wells, pools, clefts, holes, lakes, morasses, ponds, tanks, marshes, rains, rime, streams, rivers and ocean*'[56] shows that offerings are made in order to recognise the gift of water on the earth. '*Waters are healing and they strengthen one to see great joy*'[57] gives an insight into the healing powers of water. Thus it can be said that water being considered as one of the most precious gifts that has been bestowed upon us by nature, its position at

---

[48] http://www.huffingtonpost.com/entry/10-hindu-environmental-te_b_846245.html?section=india, last accessed on 03/06/2016.

[49] 'There was only water in the beginning'. Brihadaranyaka Upanishad V-5-1.

[50] Rig Veda Samhita I-10-9.

[51] Atharva Veda Samhita III-13-5.

[52] VajasaneyaSamhita IV-2.

[53] Atharva Veda Samhita X-5-24.

[54] *Id.*,XII-2-40.

[55] Rig Veda Samhita VI-39-5.

[56] Taittiriya Samhita VII-4-13.

[57] *Id.*,VII-4-19.

the pinnacle, has justified the various methods and exhortations that have been made for its conservation.

Since Hinduism believes a lot in the cleansing of both physical as well as mental attributes, water plays a great role in this respect. As a result, water is revered and protected. It is as a result of such a thought process that almost all holy places related to Hinduism are found on the banks of rivers. Water is considered by Hindus as a purifier, life-giver and destroyer of evil. Milk and water are symbols of fertility according to Hinduism. The presence of temple tanks nearby Hindu temples has great religious/spiritual significance. It is believed that a dip in the waters would cleanse one's body and soul. It is to be noted that since such beliefs are in vogue, the tanks are kept clean. Water is used as the cleansing agent for Poojas and all rituals that are associated with it.

A divine status has been accorded to rivers as far as India is concerned due to the great part they had played in influencing Indian civilisation. The river basins of Ganga and Yamuna were the seats of ancient Indian civilisation and as a result have been worshipped from time immemorial. Ganga is known for her purity and divinity, whereas Yamuna is known for her devotion. Such a status being attributed to the rivers would automatically ensure that they are taken care of. Even though the present condition of these rivers is nowhere near the state we would like it to be, it has to be kept in mind that such a status being conferred upon these rivers would definitely help us protect these water sources.

The age-old Mohenjo-daro and the Indus Valley civilisations were among the first of the civilisations that evolved in the Indian Peninsula. They too were hydraulic civilisations and akin to all great contemporary civilisations and were set up close to water sources.

While trying to understand the concept of Hinduism that was prevalent in ancient India, it is noteworthy that it was not considered a religion or a philosophy but as a way of life. The very basis of this concept was the expression of one form of the universal truth.

With the advent of the Vedic society, it can be seen that specific and specialised legal norms emerged,[58] namely, Vedas, tradition, good custom and inner contentment.[59]

Ancient Indian law took great strides in ensuring that water as a resource was protected. Through laws that were laid down and through royal declarations, such steps were taken. More about the practice and the procedures of how such laws were put into practice and how far it has helped evolve comprehensive law would be dealt with in the following parts of this work.

[58] *Infra* n. 60.

[59] Iqbal Ahmed Siddiqui, *History of Water Laws in India*, in Chatrapathi Singh (Ed), Water Law in India, ILI, 1992, at p. 290.

Administrative Architecture

While taking into consideration the advent of Indian laws as regards water, it is quite clear that the laws that were present in ancient times were based on religion and ethics.[60]

While in earlier times, the King had the prerogative and the absolute right, but with the passage of time, the same changed and evolved into a better system wherein the rights were community based.

During the reign of Chandragupta Maurya, there was a specialised 'in charge' for administration called the *Agronomoi* with duties to supervise irrigation. The officer was supposed to supervise the management of rivers, inspect sluices through which water was let out and the like. The state had the responsibility to construct tanks, canals, wells and other water sources.

Another important feature that can be found in South India is that a different system was prevalent. Herein, a special land called 'Eripatti' or 'tank land' was donated by individuals to the state for common use.[61]

Community Management

Community management of water resources has played a great role in protecting the resource. While it can be seen that there have been quite a lot of incidents wherein communities have involved themselves in protecting water and water resources, the methods and ways resorted to them would definitely help decipher their importance and how they can be put to use in the present times.

Water Legislations

When we consider water law as part of ancient times, it is quite interesting to note that there were several aspects to the same. With the advent of the mediaeval period, India saw the growth of Islamic principles of law relating to water, and it was treated as a common property resource. It was considered to be a birth right of all.[62] It can be said that mediaeval India had a system that coexisted, adjusted and assimilated different values.[63]

With the advent of the British, community-based systems suffered a lot, and state intervention increased exponentially. They tried to restructure the policies and pro-

[60] For more, see Robert Lingat, The Classical Law of India, Thompson Publishers, 1973.

[61] Romila Thapar, A History of India, Penguin Publishers, (New Delhi, 1977) at p. 184.

[62] Dante Caponera, Water Laws in Muslim Countries, 20 FAO 11 (1974).

[63] *Supra* n 59 at p. 294.

grammes based on their colonial logic. It can be seen that colonial legislations, especially the Land Acquisition Act, 1894, helped the British to acquire land rights as well as water rights.

It is also to be noted that the British, in order to promote agriculture, deforested large areas and slowly took over the ownership of those areas. This was followed by the taking over of the ownership of community irrigation. Surface water was henceforth controlled by the British, and they regulated the use of the same by a gradual introduction of common law principles. They put forward the argument that the landowner had the right to access water.

The Easements Act of 1882 contained the common law principles and evolved over time. There was also a plethora of laws that were enacted which ensured the protection and maintenance of embankments and which gave the power to acquire lands for embankments[64] and for the collector to put into practice such laws.

Laws were enacted aplenty, which regulated the use of canals, levied taxes on users and promoted the conservation of rivers and waterbodies. One of the main features was that local practices prevalent in villages were recognised and protected by law.

One of the earliest regulations that came into force was the Bengal Regulation VI of 1819. This empowered the government to invade rights of private ferry by establishing a public ferry. Later on, this was extended to Punjab too. The Embankment Regulation, 1829,[65] and the Bengal Embankment Act, 1855, were enacted so as to ensure a better protection and supervision of embankments. Modifications of the same were made in 1866, 1873 and 1882. The 1882 Act gave wider powers to the statutory authorities concerned apart from their usual powers of managing embankments and water courses. Certain acts that interfered in the embankment system were made punishable too.[66]

It is also to be kept in mind that the British created the laws relating to irrigation for India. The two Irrigation Codes, namely, the Bengal Irrigation Act, 1876, and the Bombay Irrigation Act, 1879, were prevalent during this period. The arrangement was in such a way that all water for use would be bought from the EIC by the Secretary of State. He would be handed over the money due from the users. Later on, the company sold the irrigation works to the government.[67]

As far as the legislations concerning canals and navigation were concerned, the Bengal Canals Act, 1864, and the Northern India Canal and Drainage Act, 1873, were relevant. The latter stated that 'the provincial government is entitled to use and control, for public purposes, the water of all rivers and streams flowing in natural channels and of all lakes and other natural collections of still water'.[68] It sought to

[64] The collector was given the power to ensure that such laws were implemented.

[65] Bengal Regulation 11 of 1829.

[66] S. 76, Bengal Embankment Act, 1888.

[67] This was done vide Act VI of 1869 of Bengal Council.

[68] Preamble, The Northern India Canal and Drainage Act, 1873, available at http://punjabrevenue.nic.in/canal_drain_act1.htm, last accessed on 03/06/2016.

regulate irrigation, navigation and drainage in Northern India. This act assumes significance because it brought into the picture the right of the government to control the use of water for a larger good.

Apart from the above, the Madras Rivers Conservancy Act, 1884, provided for conservancy of rivers; the Jharia Water Supply Act, 1914, dealt with drinking water supply; and the UP Municipalities Act, 1916, empowered local bodies to regulate the use of water within their jurisdiction.

The Madhya Pradesh Irrigation Act of 1931[69] provided that 'all rights in the water of any river, natural stream or natural drainage channel, natural lake or other natural collection of water shall vest in the government'.

Likewise, there have been a plethora of laws that dealt with the subject before the advent of the Constitution of India. That being treated as the cut-off point, the author would not be discussing the various codes and laws that prevailed after the enactment of the constitution.

Grievance Redressal, Conflict Resolution and Justice Delivery

The Dharmasutra, which can be treated as a text on law, of Apastambas states that any person taking over the property of another shall be reprimanded. However, it also states that if the same has been done intentionally, then the garments of the person shall be taken away.[70]

However, it is quite interesting to note here that the King was given the power to take or receive water from any place. The same would not be considered a theft as he enjoyed a legal immunity by virtue of his being the sovereign.[71]

There are also instances wherein you can see that the embankments were given protection. This could be due to the fact that these were helpful not only in protecting the kingdom from the harmful effects of water but also were used for storing water. Based on the seriousness of the issue of protecting the embankments, it can be seen that severe punishments including capital punishment were prescribed for destroying embankments. Manu had laid down that whoever destroyed an embankment of a tank would be drowned or beheaded. Capital punishment was provided for by the Vishnu Sutras too.[72]

It can also be seen that heavy fines were also imposed for unauthorised use of tanks and destruction of embankments.[73]

[69] Available at http://www.ielrc.org/content/e3102.pdf, last accessed on 03/06/2016.

[70] Jolly J, Naradiya Dharmashastra, Delhi 1981, at pp. 50–51.

[71] *Ibid* at p. 114.

[72] Max Muller (Ed.), Sacred Books of the East, Vol. IX, Motilal Bansaridas, Delhi, (1965), at p. 279.

[73] *Ibid.*

Comparisons with Other Systems

As a way of comparison, a few other systems have been looked into, so as to really comprehend the growth of legal principles as well as to understand the extent of such principles.

Egyptian Civilisation

The first records which helped in identifying water resources being put into use in ancient Egypt, though very little is known of them, have been identified. References to a codification by name Horemheb at Karnak have been made in the literature, but sadly none of them survived. Some historians say that some reliable information is present which throws light on the fact that ancient Egypt might have had a set of rules that dealt exclusively with water. Nile, the sole provider of water in the arid desert, was sought to be protected. During the time of the first two dynasties (3400–2980 BC) itself, there was a 'Constructor of the King' who was in charge of public works. The Nile was measured frequently and was also recorded.

It is also to be noted that the administration of water used to be overlooked by the Constructor of the King. This in the later stages transformed into two departments during the rule of the third dynasty (2980–2900 BC). It was then headed by the Master of Canals and the Master of Lakes. A further addition of Master of Water Castles was made during the Fifth Dynasty (2750–2625 BC).[74]

During the later periods (2650–300 BC), the district governors had a special category of people called 'Digger of Canals' who were responsible for the upkeep and maintenance of the canals.[75] They were also supposed to ensure the safety of people and the city whenever water reached its maximum height. They had a duty to perform in the event of any disaster as regards the same.

The King of Egypt, the Pharaoh, was considered the living God. It was up to him to grant rights to people to use land and water, and, in return, such persons who were given rights were supposed to pay him a share of the produce they grew although there was a practice of recording done by scribes, no written water regulations were there, probably due to the very fact that the dictum of the Pharaoh itself was considered a law.[76]

However, it is quite sad to note that very little is known about the land and water institutions. It has been opined that there is a strong evidence to show that a strong state control over water and land was present.[77]

---

[74] See Dante A Caponera, *Principles of Water Law and Administration*, Taylor & Francis, 2007.

[75] *Ibid.*

[76] J.A. Wilson, *The Burden of Egypt,* University of Chicago Press, Chicago, 1951, pp. 41–50.

[77] Joshua Getzler, A History of Water Rights at Common Law, Oxford University Press, Oxford, 2004, at p. 10.

Mesopotamian Civilisation

By virtue of being on the banks of the Euphrates and the Tigris, the Mesopotamian civilisation too had its share in the history of water governance. The very name means 'land between rivers'. It was considered to be the cradle of civilisation as many a civilisation took birth there and flourished.

The first law codes were probably formed in the city states of Mesopotamia. The codes of Urukagina[78] and Lipit-Ishtar[79] are two important codes that took birth during this period. However, from the parts that have been recovered, nothing relates to water governance.

The Code of Ur-Nammu

One of the earliest references to law related to water and its use can be found in the Code of Ur-Nammu.[80] Written on tablets in Sumerian language, the first copy of the code was discovered in 1952. It was found in two fragments by Mr. Samuel Kramer,[81] at Nippur, and translated by him. It was only in 1965 that a few more fragments of the same were discovered. The very significance of the same is that though we have had discoveries made of legal texts and laws prevailing, which predated the Ur-Nammu code, this was the first of a comprehensive legal text as such.

It is quite interesting to note that there was a law in this code which specified that

> *If a man flooded the field of a man with water, he shall measure out three kur of barley per iku of field.*[82]

This was perhaps the first ever law that had a direct relation to water.

Code of Hammurabi

When we talk of Mesopotamia and its laws, the first thing that comes to our mind is the Code of Hammurabi (1780 BC).[83] From the transliteration made, it can be seen that quite a large number of references are made to water. The prologue specifies

---

[78] For more, see http://en.wikipedia.org/wiki/Urukagina, last visited on Dec 10, 2013.

[79] He was the fifth ruler of the first dynasty of Isin and ruled from around 1870 BCE to 1860 BC. Some documents and royal inscriptions from his time have survived, but he is mostly known for Sumerian language hymns written in his honour, as well as a legal code written in his name. For more, see http://en.wikipedia.org/wiki/Lipit_Ishtar, last visited on Dec 10, 2013.

[80] These were written somewhere between 2100 and 2050 BC.

[81] He is one of the world's leading Assyriologists and a world-renowned expert in Sumerian history and Sumerian language.

[82] http://en.wikipedia.org/wiki/Code_of_Ur-Nammu, last visited on Dec 10, 2013.

[83] http://en.wikipedia.org/wiki/File:The_code_of_Hammurabi.pdf, last visited on Dec 10, 2013.

that he supplied water in abundance to the inhabitants[84] and also identified pastures and watering places for Lagash and Girsu.[85]

Articles 53 to 56 specifically dealt with water. The same has been discussed below.

Article 53[86] states as follows:

> *If a man neglect to strengthen his dyke and do not strengthen it, and a break be made in his dyke and the water carry away the farm-land, the man in whose dyke the break has been made shall restore the grain which he has damaged.*

Article 54[87] is a continuation of the previous article. It states that:

> *If he be not able to restore the grain, they shall sell him and his goods, and tile farmers whose grain the water has carried away shall share (the proceeds of the sale).*

Article 55[88] states that:

> *If a man opens his canal for irrigation and neglects it and the water carries away an adjacent field, he shall measure out grain on the basis of the adjacent fields.*

Article 56[89] states that:

> *If a man opens up the water and the water carries away the improvements of an adjacent field, he shall measure out ten gur of grain per gan.*

The epilogue also refers to various divine punishments meted out to those found violating these laws.[90] The wrath of God would be upon the person who tended to disobey the said rules.

From the above-mentioned laws, we can surmise that there was some sort of a rudimentary form of the concept of 'negligence' that we have in our present-day laws. The person who was to exercise care and caution as regards water use would have to make good the loss suffered by another in case of any damage caused by water.

The administrative authority in place in the city state was responsible for coming up with rules concerning the use of water and also the maintenance works that needed to be taken up, time and again. Maintenance works were supposed to be taken up as and when required, and it was the duty of landowners to carry out the same. In return, the land owners were given fishing rights in their part of the waterway. This ensured that the land owners kept the water sources in good condition.

A hierarchical structure with the King at the top was present. The duty of the King was to ensure that the customary rules as regards administration of water resources were followed and, that in cases where such rules were absent, to turn to

---

[84] *Ibid*, at p. 5.

[85] *Ibid*, at p. 7.

[86] *Ibid*, at p. 29.

[87] *Ibid*, at p. 31.

[88] *Ibid*.

[89] *Ibid*.

[90] *Ibid*, at pp. 105–106.

analogy and come up with relevant ones. Below him there was a decentralised structure which included the circumspection chief or the first civil authority and the local council. The latter included the mayor and the 'rector of the fields'.[91]

There are also references to the harmful effects of water. It is noted that the Rain God, Adad, if not satisfied, would wreak havoc. Moreover, the very basis of attaching an obligation to compensate the victim, on account of damage caused by water, to the person who was supposed to have exercised care and caution also throws light on the fact that the harmful effects of water too were taken into consideration while coming up with laws and rules.

While it is evident that water-related crimes were punished, it is quite interesting to take note of the fact that water was also used as a means of punishment. Execution by drowning in water was the most common form of punishment and was followed in many codes that were prevalent in those times.[92]

Roman Law

One of the earliest systems of law that was prevalent which had a profound influence over succeeding legal systems was the Roman Law, which has had a significant role in shaping the world's legal history.

While delving into the history of Rome, it can be found out that during the early period,[93] the sources of law were:

1. Custom (ius non scriptum)
2. King-made laws (leges regiae)

There was no reference to any laws on water and its use during this period of time.

The Republican Period[94] saw the Law of the Twelve Tables (Latin: *Leges Duodecim Tabularum* or *Duodecim Tabulae*) come into existence which was the ancient legislation that was the basis of Roman law, on which even the Roman Republic Constitution owed quite a lot to the set of rules laid down in the same. This also formed the crux of the *mos maiorum* (custom of the ancestors). This was the foremost of the law codes written by Romans. It dealt with many topics that ranged from public, private, religious, criminal and procedural rules that were enacted so as to ensure that a healthy relation existed among the people of Rome.

Even back then, the importance attached to water can be understood from the fact that there were references to water.

---

[91] *Supra* n.74.

[92] *Ibid.*

[93] This period originated from the founding of Rome in 753 BC to the beginning of the Republic in 509 BC.

[94] 509 BC to 27 BC.

In Table VIII concerning the Laws of Real Property, Law VII states that:

When rain falls upon the land of one person in such a quantity as to cause water to rise and injure the property of another, the Prætor shall appoint three arbiters for the purpose of confining the water, and providing against damage to the other party.[95]

Apart from the Twelve Tables, this period also saw the birth of many other laws that were related to the use of water.

1. Lex Agraria[96] (112 BC)

   The importance of this set of laws is that during those periods of time, the legal position attached to land had a direct relevance to the legal position given to water.

   It can be seen that a privileged position was accorded to land owners whose land was next to water courses. Being a cardinal principle of Roman water law, it influenced the origins of water rights in European legal tradition, much before the advent of modern water rights systems; some of which is even to date observed.[97]
2. Lex Mamilia Roscia (58 BC)

   A reference was made in this to impose fines on anybody who prevented the water from flowing into border ditches. The fines were to be collected by the municipality.

   As discussed earlier, springs and wells were considered arising out of land, and the status of water depended very much on the status that was accorded to land. The public land where water was available in whatever nature was deemed to be public property as well. Similarly all forms of water, if they fell within private lands, were deemed to be private in nature.

Property Aspect of Water Under Roman Law

It can also be seen that there were two categories of property, namely:

(a) Material things over which there was an unrestricted right to control[98]
(b) Property beyond private control and ownership[99]

Roman law treated water as part of the second maxim, i.e. one over which there was no private control. However, it is to be understood that such common property resources were considered to come under state control so as to ensure that the same was not over-exploited and misused.[100]

---

[95] http://www.constitution.org/sps/sps01_1.htm, last accessed on 03/06/2016.

[96] https://en.wikipedia.org/wiki/Lex_Agraria, last accessed on 03/06/2016.

[97] Historical Water Rights: The Lost Connection between Land Tenure Rights and Water Rights, available at http://www.fao.org/docrep/007/y5692e/y5692e05.htm, visited on 20/12/2013.

[98] As portrayed by the Latin Maxim, 'dominium ex jure Quiritium'.

[99] As portrayed by the Latin Maxim, 'res extra commercium'.

[100] Rajiv Khare, The Journey of Water Resources Law and its Management: Doom or Dawn?, The ICFAI University Press, Vol V, No. 3, 2006.

Justinian, the great Roman Emperor, classified water, air and the sea as *res communes,* i.e. those available to all. Rights over these were termed as *ius naturale*.[101] For a resource to be classified under that head, it should be plentiful as well as pure, and water sure had those qualities.

Islamic Law

Islamic law or the Shari'a governs the whole of the Islamic world. The followers of this religion keenly follow the tenets that have been laid down. The growth of Shari'a[102] law has also contributed much towards the protection of the environment and casts a positive duty upon the followers to follow the tenets of Islam and play their role in preserving the earth and its produces. As a result of the growth of Shari'a law, there was an emergence of various legislative principles[103]:

- Allah is the sole owner of the earth and everything in it. People hold land as a usufruct, i.e. for its utility value only. There is a restricted right to public property.
- Abuse of rights is prohibited and penalised.
- There are rights to the benefits derived from natural resources held in common.
- Scarce resource utilisation is controlled.
- The common welfare is protected.
- Benefits are protected and detriments are either reduced or eliminated.

The growth of Shari'a law also gave rise to various institutions that facilitated environment protection.[104]

It can be seen that an agency known as the *Hisba* was established in the Islamic states right from the very early times. The main function was to defend the people by way of encouraging the establishment of good and by forbidding wrongdoing. A learned jurist known as *Muhtasib* headed this.[105]

Now that man can use, utilise and harness the natural resources, it is quite natural that corresponding to these rights, there are certain duties that have to be cast upon him to ensure that the rights are used in a correct way. This is to ensure that man, after using the resources, does everything in his power to ensure that he conserves the resources qualitatively and quantitatively. There is another similar line of thought which explains that Allah created this world and all that it encompasses to ensure that man realises the objectives of contemplation, worship, inhabitation, construction, sustainable utilisation and enjoyment and appreciation of beauty. Due to

---

[101] Institutes of Justinian, Book 11 Title 1.

[102] The Arabic word translates as 'track' or 'road'.

[103] Fazlun M Khalid, 'Islam and the Environment' in Mr Peter Timmerman (Ed.), *Encyclopedia of Global Environmental Change*, John Wiley & Sons Ltd, Chichester, Vol.5, 2002, p. 335.

[104] *Ibid.*

[105] *Ibid.*

such a concept, it is clear that man has no right to exploit the resources solely for his benefit.[106]

Thus it can be said that the attitude of Islam towards the environment and its resources is based on the prohibition of abuse and the notion of sustainable development. This idea is what was contemplated in bringing unto life on this earth and allowing it to flourish through agriculture, cultivation and construction.[107] From all these, it can be understood that both conservation and development of the environment is for the universal good of all.

Each thing and every creature in the universe, whether known to man or not, according to the Quran performs two major functions[108]:

- A religious function in so far as it evidences the Maker's presence and infinite wisdom, power and grace
- A social function in the service of man and other created beings

Each creature performs its preordained role, and this helps in maintaining the dynamic balance. It can be seen that the problems caused by over-exploitation, abuse, misuse, destruction and pollution of natural resources are all contraventions against the scheme of things of the Almighty. It is an obligatory duty to protect all natural resources from abuse. Although it can be said that the societal functions of all things are important, the comprehensive legal basis for conserving the environment is the fact that all created beings are considered as symbols of the Creator.

Benefits that have been received by one from Allah as natural resources should be respected by a believer, commands Islam. A thrift use of the available resources is what is expected, and an extravagant life with all the luxuries is to be avoided at all costs. The basic idea derived from the Quran—consume from the eatables and healthy drinks but do not indulge in excess—is the key to a healthy individual and society.[109]

The teachings profess the idea that the creation should be respected for the very fact that it is so valued and premeditated for such a high ideal. It also conveys the idea that we humans can never be able to understand all the beneficial functions of all things that are present on this earth. In this very respect, it is said that if we take into consideration environmental benefits of only humans, it would distort the balance created by Allah in this world. What has to be understood is that we are to not only take into account the consequences of our actions on us alone but also on the whole of the earth as such. The concept of haves and have-nots as visualised in Islam identifies that such a situation would naturally arise if the vital resources of

[106] The Prophet declared: '*If any Muslim plants a tree or sows a field, and a human, bird or animal eats from it, it shall be reckoned as charity from him*'. Saheeh Al-Bukhari, Saheeh Muslim.

[107] Quran 11:61 '*…It is He Who has produced you from the Earth and settled you therein…*'.

[108] Dr. A. Bagader*et.al.*(Eds.), 'Environmental Protection in Islam- A General Introduction', available at http://www.islamreligion.com /articles/ 307/viewall/, accessed on 20/10/2013.

[109] Quran 6:141.

the earth are not shared equally. This is the basis on which the concept of *zakath*[110] and the prohibition of *riba*[111] come into force.

Islam forbids damage of all forms and kinds.[112] One of the fundamental principles of Islamic law is the prophetic declaration: '*There shall be no damage and no infliction of damage*'.[113] 'Prevention is better than cure' is the idea that has been put in practice here. An important juristic rule in Islamic law states 'The averting of harm takes precedence over the acquisition of benefits'.[114] Thus, to do everything by causing minimal harm is one of the cardinal principles which can be found in Islam.

Principles Relating to Water

Since Islam started with a desert community, it had a special concern for the environment as is quite evident that water resources are scarce in a desert. This is one of the main reasons as to why conservation of resources such as water and vegetation was given much importance. Scarcity of resources made them realise the true value of them, and they had no other way but to make the most out of those resources available to them. This fostered and called for the optimal use of resources as mandated by the religion and its principles.

Laws relating to water form an integral part of Islamic law for the reasons mentioned above. It purports to make it clear that every member of the community has access to water. God has made water the basis and origin of life.[115] Every single being which has life, whether it be plants, animals or even man, all depend on water for their existence and the continuation of their lives. '*Verily... in the rain that God sends down from Heaven, thereby giving life to the Earth after its death...*'.[116] '*It is He who sends down water from the sky; and thereby We have brought forth the plants of every kind...*'.[117]

It is apparent that God has directed man to realise the value of water as is evident in the verse '*Have you seen the water which you drink? Was it you who sent it down from the rain cloud, or did We send it? Were it Our will, We could have made it bitter; why then do you not give thanks*'?[118] God also ensures that man remembers that

[110] Alms-giving to the poor.

[111] Usury/interest.

[112] Akrum Helfaya, Amr Kotb, Rasha Hanafi, *Qur'anic Ethics for Environmental Responsibility: Implications for Business Practice,* J Bus Ethics DOI 10.1007/s10551-016-3195-6, available at http://link.springer.com/ article/10.1007%2Fs10551-016-3195-6, accessed on 03/06/2016.

[113] *Ibid.* See also Al-Hakim.

[114] See maidenheadinterfaith.org.uk/download/exhibit.environment.pdf, last accessed on 20/02/2014.

[115] Quran 21:30 God says: '*We made from water every living thing…*'.

[116] *Id.*, 2:164.

[117] *Id.*, 6:99.

[118] *Id.*, 56:68–70.

it is a gift from Him when he asks them '*Say: Have you considered, if your water were one morning to have seeped away, who then could bring you clear-flowing water*' ?[119]

Apart from these functions that have been clearly laid down, there is a sociorelious function too attached. Water is considered to be an element which purifies the body and clothing from all dirt, impurities and defilement.[120] This is to ensure that man encounters God clean and pure. Many other functions of water have been made evident with the fact that water has been shown as a habitat for many created beings. The importance that it holds can also be understood from the following verses. '*It is He Who has made the sea of service, that you may eat thereof flesh that is fresh and tender, and that you may bring forth from it ornaments to wear, and you see the ships therein that plough the waves, that you may seek of His bounty*'[121] and '*Lawful to you is the pursuit of water-game and its use for food a provision for you, and for those who travel...*'.[122]

Conservation of water is obligatory in Islam as there is a concept that whatever is essential to fulfil the crucial responsibility of preserving life is in itself obligatory. Due to the utmost importance that water has in the lives of all living things, God has made its use as the common right of all living beings. God commanded with regard to the people of Thamud and their camel, '*And tell them that the water shall be shared between them...*',[123] and the prophet said: '*Muslims are to share in these three things: water, pasture, and fire*'.[124]

A special care has to be given to ensure that extravagance in the use of water is not there. An anecdote refers to the prophet asking his companion Sa'd, who was washing himself for prayer: '*What is this wastage, O Sa' d*'? *Is there wastage even in washing for prayer? asked Sa' d; and He said, 'Yes, even if you are by a flowing river*'![125]

Water Laws

The basic principles, discussed above, have exerted a great influence on people in countries which follow Islamic law. It is for this very fact that in such countries, while enacting a water law, due consideration is given to all those principles laid down in the scriptures.

---

[119] *Id.*, 67:30.

[120] *Id.*, 8:11 '*And He caused rain to descend on you from heaven to cleanse you therewith…*'.

[121] *Id.*, 16:14.

[122] *Id.*, 5:96.

[123] *Id.*, 54:28.

[124] http://www.islamreligion.com/articles/307/viewall/environmental-protection-in-islam/, last accessed on 03/06/2016. Also see Abu-Dawud, IbnMajah and al-Khallal.

[125] http://www.the-faith.com/featured/water-and-air/, last accessed on 03/06/2016. See also Ibn Majah.

Before Prophet Muhammed, there were no concrete water regulations. Based on the nature of life that the people of the Middle East led, it was quite common for them to settle near waterbodies, as water was quite scarce there. It was based on the teachings of the prophet that the need to conserve water and use it judiciously began to surface. Water was portrayed as a gift and was treated to be a common property of the people. The prophet prohibited the selling of it, too. Another important feature to be noted is that the prophet took into account the ownership of wells, water sources and canals as well and put forth that it was unlawful to dig a new well in the neighbouring land as it would reduce the quantity of water in the already existing ones.

The first codified laws on water rights were issued in countries of North Africa. At the very beginning, they consisted of ordinances, orders and decrees promulgated by the Governors in the states. Certain cardinal principles were followed.

1. The state had complete control over water and water resources.
2. Water use except for drinking and animal watering purposes was brought under the government control.
3. To survey and recognise water rights, water commissions were set up.
4. There was a system of keeping track of land and water rights in specified registers.

The Code of the Ottoman Turks

The late nineteenth century and early twentieth century witnessed the effort to codify the Shari'a law. The Ottomans were the first to have a civil code during this time called the Mecelle or Mejelle.[126] Ahmed Cevdet Pasha headed the commission that prepared this code which is spread over 16 volumes and has 1851 articles. Compilation of the same was done from 1869 to 1876, and the same was enforced in the year 1877.

The Hanafi legal tradition enjoyed an official status in the empire. The code followed a European frame due to this fact. Even after the First World War which resulted in the fall of the Ottoman Empire, this code continued to exert its impact on most of the successor states.[127] This was due to the fact that the code was effective, coherent and difficult to dislodge. It remained in force till the twentieth century AD in many Islamic countries.[128]

Water rights have been detailed in the same, and references to it are reproduced.

[126] The Mejelle (Majallah el-Ahkam-i-Adilya) was elaborated between 1869 and 1876 as part of the legislative purpose of the tanzimat and based on the Hanafi law of fiqh.

[127] Egypt however was a notable exception. The system was never in force there.

[128] For more, see http://en.wikipedia.org/wiki/Mecelle, visited on 20/12/2013. The code can be viewed at http://www.iium.edu.my/deed/lawbase/al_majalle/index.html, last accessed on 03/06/2016.

As far as the right of taking water is concerned, it is the right of taking a clearly defined and ascertained share of water from a river.[129] The common property resource aspect was put into play here.

In what can be said to be a precursor to the easmentary rights that we have now, the sale of a right of way, of a right of taking water and of a right of flow attached to land and of water attached to canals is valid.[130]

The principle of precaution is detailed in the law which states that if A, without any justification, cuts off water in B's field or garden and if the crops and plantations dry up and are destroyed or if A lets the water overflow into the garden of another and swamps his crops, causing them to be destroyed, A must make good the loss.[131]

A few articles put forth certain principles with certain elements that would be of interest in this paper—a joint owner of a servitude is a person who shares with another his right over property held in absolute ownership, such as a share in water or a share in road[132]; a private right of taking water is a right of taking water from some flowing water reserved for a limited number of persons. But the right of taking water from rivers used by the public does not belong to this class[133]; if the owner of a garden possessing a private right of taking water sells such garden without the right of taking water, those persons who share in the right of taking water cannot claim a right of pre-emption.[134] The same principle is applied in the case of a private road. A right of taking water is preferred to a right of way.[135] Therefore, if upon the sale of a garden in respect of which one person is the joint owner of a private right of taking water and another of a private right of way attaching thereto, the owner of the right of taking water is preferred to the owner of the right of way.[136]

---

129 Article 143, The Mejelle, available at http://www.iium.edu.my/deed/lawbase/al_majalle/index.html, last accessed on 03/06/2016.

130 Article 216, The Mejelle, available at http://www.iium.edu.my/deed/lawbase/al_majalle/index.html, last accessed on 03/06/2016.

131 Article 922 (2), The Mejelle, available at http://www.iium.edu.my/deed/lawbase/al_majalle/index.html, last accessed on 03/06/2016.

132 Article 954, The Mejelle, available at http://www.iium.edu.my/deed/lawbase/al_majalle/index.html, last accessed on 03/06/2016.

133 Article 955, The Mejelle, available at http://www.iium.edu.my/deed/lawbase/al_majalle/index.html, last accessed on 03/06/2016.

134 Article 1015, The Mejelle, available at http://www.iium.edu.my/deed/lawbase/al_majalle/index.html, last accessed on 03/06/2016.

135 Article 1016, The Mejelle, available at http://www.iium.edu.my/deed/lawbase/al_majalle/index.html, last accessed on 03/06/2016.

136 *Ibid.*

Joint Ownership

Book X details about joint ownership. It has been described as follows: *Joint ownership consists of a thing itself belonging absolutely to more than one person, so that such persons enjoy a special position in relation to such a thing. It is also customary to apply this expression to a contract whereby the state of joint ownership is brought about, and is used in this sense in technical legal phraseology. Consequently, joint ownership is generally divided into two classes. The first consists of joint ownership of property held in absolute ownership brought about by one of the modes of acquiring property, such as purchase, or the acceptance of a gift. The second consists of joint ownership as a result of contract brought about by the offer and acceptance of the joint owners, the details concerning both of which are dealt with in the relevant Chapters. Another class consists of gratuitous joint ownership which is brought about by the joint acquisition of ownership by the public of things which are free and themselves belong absolutely to no particular person, such as water.*[137]

Judicial Use of Water

Further, a description is given which details as to how water is to be managed and used judiciously. All these invariably put forward the fact that water, treated as a resource needing much protection and preservation, owing to the desert conditions and scarcity, was taken care of well.

Certain definitions are put forward in the sections, for example, by water channels, what is meant are pipes and underground channels for conducting water,[138] and by dam, what is meant is any boundary or water dam and the sides of any water channel.[139]

The rudiments of reasonable care, making good the loss suffered by another and strict liability principle can be found in Article 1212, which states that any person constructing a cesspit or a sewer near a well belonging to some other person, and contaminates the water thereof, may be made to remove the injury. It further states that if it is impossible to remove the injury, the person may be made to close up the cesspit or sewer. Further, if any person constructs a sewer near a water channel, and the dirty water from such sewer flows into the channel and causes great injury thereto, and no other way can be found to remove such injury than by closing it, the sewer shall be closed.[140]

[137] Article 1045, The Mejelle, available at http://www.iium.edu.my/deed/lawbase/al_majalle/index.html, last accessed on 03/06/2016.

[138] Article 1049, The Mejelle, available at http://www.iium.edu.my/deed/lawbase/al_majalle/index.html, last accessed on 03/06/2016.

[139] Article 1050, The Mejelle, available at http://www.iium.edu.my/deed/lawbase/al_majalle/index.html, last accessed on 03/06/2016.

[140] Article 1212 , The Mejelle, available at http://www.iium.edu.my/deed/lawbase/al_majalle/index.html, last accessed on 03/06/2016.

Joint Property

Water, grass and fire are free. The public are joint owners of these three things and have the right over the same.[141] Water flowing underground is not the absolute property of any person which is the underlying principle laid down in Article 1235. The aspect of community rights can be understood from this. This article strengthens the idea that not only the person who owns the land but also other persons, who might not necessarily own land, too, has an equal right to water. The idea that even underground water is for all ensures that the sole possession or privilege is not accorded to a few so as to exclude the rest.

The public right over water is further accentuated by the law that states that wells which have not been constructed by the labour of any particular person, the benefit of which may be enjoyed by the public, are the jointly owned and free property of the public,[142] and seas and large lakes are free.[143]

By virtue of Article 1238, rivers which belong to the state and are not the property owned by any person are those rivers the beds of which do not pass through the property of a group of persons owned in absolute ownership and are free.[144]

Article 1239 states that rivers which are the property of individuals owned in absolute ownership, that is to say, rivers which flow through the property of persons owned in absolute ownership are of two categories.[145]

> The first category consists of rivers the waters of which are divided between the joint owners of the land through they flow, but is not completely exhausted and continues its course through vacant lands which is free to public. Rivers of this class are called public rivers, since they are at the disposal of the public. No right of pre-emption is attached to these rivers.
>
> The second category consists of private rivers, the waters of which are divided between the land belonging to a limited number of persons, and which, upon arriving at the limits of such land, disappear and do not flow in vacant land. A right of pre-emption is attached to such land.[146]

The next sections deal with acquiring the ownership of those things which are free.

By virtue of Article 1248, there are three means of acquiring absolute ownership. The first consists of the transfer of property held in absolute ownership from one owner to another, such as sale or gift. The second consists of one person succeeding

[141] Article 1234, The Mejelle, available at http://www.iium.edu.my/deed/lawbase/al_majalle/index.html, last accessed on 03/06/2016.

[142] Article 1236, The Mejelle, available at http://www.iium.edu.my/deed/lawbase/al_majalle/index.html, last accessed on 03/06/2016.

[143] Article 1237, The Mejelle, available at http://www.iium.edu.my/deed/lawbase/al_majalle/index.html, last accessed on 03/06/2016.

[144] The Mejelle , available at http://www.iium.edu.my/deed/lawbase/al_majalle/index.html, last accessed on 03/06/2016.

[145] *Ibid.*

[146] Article 1239, The Mejelle, available at http://www.iium.edu.my/deed/lawbase/al_majalle/index.html, last accessed on 03/06/2016.

another, such as inheritance. The third consists of obtaining a thing which is free and which has no owner. The latter is either actual, as where someone in fact appropriates such thing, or constructive, as where someone puts out a receptacle to collect rainwater or sets a trap to catch game.[147]

Ownership, the ways as to how it is fixed, can be understood from a few sections which say that any person, who obtains possession of a thing which is free, becomes the independent owner thereof.[148] It is also interesting to note that taking possession of a thing must be coupled with intention.[149] In furtherance to this concept, it is said that if any person puts out a receptacle with the object of catching rainwater, the rainwater caught in the receptacle becomes that person's property. Again, water is collected in a receptacle not intentionally put in any particular place and does not become the property of the owner thereof. Any other person may take it and consume it.[150]

While dealing with the aspect of taking possession of water, the flow thereof must be interrupted.[151] Consequently, possession cannot be taken of water from a well which oozes out from the sides thereof and if a person takes and uses the water, he is not liable to make good the loss thereof, even though the owner has not made a free gift thereof for consumption. Similarly, possession cannot be taken of water and the flow of which is regulated, that is to say, water which leaves one side of a tank in the same quantity as it enters the other side.[152]

Rights of Taking Water and Right of Drinking Water

Water is considered to be the resource that is to be put to use very judiciously, and the rights of taking water and drinking water have been specifically laid down in the code.

By watering, what is meant is taking one's turn in making use of water to water crops and animals.[153] The very fact that both crops as well as animals too need water and are given it shows the consideration that was given, not only to humans but also every life form. This is further strengthened by the fact that is specifically laid down

[147] Article 1248, The Mejelle, available at http://www.iium.edu.my/deed/lawbase/al_majalle/index.html, last accessed on 03/06/2016.

[148] Article 1249, The Mejelle, available at http://www.iium.edu.my/deed/lawbase/al_majalle/index.html, last accessed on 03/06/2016.

[149] Article 1250, The Mejelle, available at http://www.iium.edu.my/deed/lawbase/al_majalle/index.html, last accessed on 03/06/2016.

[150] *Ibid.*

[151] Article 1251, The Mejelle, available at http://www.iium.edu.my/deed/lawbase/al_majalle/index.html, last accessed on 03/06/2016.

[152] *Ibid.*

[153] Article 1262, The Mejelle, available at http://www.iium.edu.my/deed/lawbase/al_majalle/index.html, last accessed on 03/06/2016.

that the right of drinking consists of the right of drinking water[154] and that any person may make use of air and light and of seas and big lakes.[155]

Care and caution too need be exercised while making use of the resource. This is accentuated by the fact that if suitable care and caution are not exercised and the resource causes harm, the use of the same is to be curtailed and stopped. This concept can be understood from the fact that any person may water his lands from rivers which are not owned in absolute ownership by any particular person and, in order to irrigate them and to construct mills, may open a canal or water channels, provided that he does not thereby inflict injury on any other person. Consequently, if the water overflows and causes injury to the public, the water of the river is entirely cut-off or boats cannot be navigated, such injury must be stopped.[156]

Water, treated as a common resource, has to be made available for all, and it is laid down that all persons and animals have a right of drinking water; the possession of which has not been taken over by any other person.[157]

However, there are certain restrictions placed on privately owned water. The right of taking water from rivers privately owned, that is to say, the courses of which are privately owned, belongs to the owners thereof. Other persons have a right of drinking there from. Consequently, no person may, without permission, water his land from a river which is appropriated to a group of persons. He may, however, drink water there from, since he has a right of drinking water. He may also water his animals provided there is no danger of destroying the same. He may also bring water to his house or his garden by means of jugs or buckets.[158] All these lee ways that are provided for take into consideration the fact that water which is a public resource should be taken care of well, and everyone has a right to use it judiciously.

The concept of sharing of resources that are scarce can be noted in Article1268, which ensures that everyone gets an access to water. Herein, it is stated that if any person having in his property, which he owns in absolute ownership, a tank, a well or a river from which water alternatively enters and leaves, he may prevent any person who wishes to drink water from entering his property. However, if there is no free water to be had in the neighbourhood, the owner of the property is obliged either to draw off water or to give such person permission to enter his property and take it.[159]

---

[154] Article 1263, The Mejelle, available at http://www.iium.edu.my/deed/lawbase/al_majalle/index.html, last accessed on 03/06/2016.

[155] Article 1264, The Mejelle, available at http://www.iium.edu.my/deed/lawbase/al_majalle/index.html, last accessed on 03/06/2016.

[156] Article 1265, The Mejelle, available at http://www.iium.edu.my/deed/lawbase/al_majalle/index.html, last accessed on 03/06/2016.

[157] Article 1266, The Mejelle, available at http://www.iium.edu.my/deed/lawbase/al_majalle/index.html, last accessed on 03/06/2016.

[158] Article 1267, The Mejelle, available at http://www.iium.edu.my/deed/lawbase/al_majalle/index.html, last accessed on 03/06/2016.

[159] Article 1268, The Mejelle, available at http://www.iium.edu.my/deed/lawbase/al_majalle/index.html, last accessed on 03/06/2016.

In cases where the owner does not allow, then such person has the right of entering and taking it, subject, however, to the fact that no injury is caused. This ensures that the resource is used judiciously and shared among all who are in need.

To ensure that a person does not have a prerogative and does not use more than what he is supposed to use, it is stated that a joint owner of a river may not open up another there from, unless he has obtained the permission of the other joint owners. He is also not given the power to alter the old established order in which he has his right of taking water. Nor may he divert his share of the water from such a river on to other land not enjoying a right of taking water. If the other joint owners agree to such things, either they or their heirs may denounce such an agreement at any subsequent date.[160]

Cleaning and Improvement of Rivers and Water Courses

Cleaning and improvement of water resources is accorded much needed significance and priority. It is stated that the cleaning and improvement of rivers which do not belong to any particular person in absolute ownership is incumbent upon the Treasury. However, it is also stated that if it is not in the power of Treasury to do so, the public may be forced to do so.[161] This ensures that the public are always kept in the loop and for the benefit that they enjoy, they are also cast upon a duty to protect it so as to ensure that the same would be protected and in good condition.

It is laid down that the cleaning of rivers jointly owned in absolute ownership is incumbent upon the owners thereof, that is to say, upon those who have the right of taking water there from. The owners with a right of drinking water may not be called upon to share the expenses of cleaning and improvement.[162] This ensures that the one with greater powers is given greater responsibilities.

The importance of keeping the water sources clean is given a high degree of prominence. It is noted that if some of the owners with a right of taking water from a jointly owned river desire to clean such river and the others refuse to do so, the persons who refuse will be made to clean such river jointly with the others, if it is a public river. If it is a private river, those persons who wish to clean it may, by order of the court, proceed to do so and may also prevent those who refuse from making use of the river until such time as they have paid the amount which falls to their share of the expenses.[163] This ensures that those who are enjoying the resource

[160] Article 1269, The Mejelle, available at http://www.iium.edu.my/deed/lawbase/al_majalle/index.html, last accessed on 03/06/2016.

[161] Article 1321, The Mejelle, available at http://www.iium.edu.my/deed/lawbase/al_majalle/index.html, last accessed on 03/06/2016.

[162] Article 1322, The Mejelle, available at http://www.iium.edu.my/deed/lawbase/al_majalle/index.html, last accessed on 03/06/2016.

[163] Article 1323, The Mejelle, available at http://www.iium.edu.my/deed/lawbase/al_majalle/index.html, last accessed on 03/06/2016.

would definitely keep it clean and would take adequate steps for the same. Not allowing them to use the facility is the best way of penalising and ensuring cooperation in keeping it neat and clean.

The importance attached to the duty to keep the resources clean can be understood from the fact that all the owners who have a right of taking water are supposed to keep the sources clean. If they refuse to do so, they may be forced to do so, if it is a public river, but not if it is a private river.[164] This is to ensure that others too access the same.

The importance of water as a resource and making it available to all can be found in the section that details that if any person owns land on the banks of a public river, whether such river is in absolute ownership or not, and there is no other way for satisfying such needs as drinking water or improving the river, the public may pass over such land, and the owner cannot prevent them from doing so.[165]

Public participation is the key, and it is ensured in such a way and manner that all persons who use the resource are made to take part in cleaning the same. Expenses connected with the cleaning and improvement of a jointly owned river begin from the upper strata. First of all, the whole of the joint owners must share therein, beginning with the joint owner whose land comes last, the reason being that disadvantage is an obligation accompanying enjoyment.[166]

Dealing with Limitations

The principle of limitation was also in prevalence and was laid down that actions relating to a private road, to a right of flow and to a right of taking water, when relating to real property held in absolute ownership, shall not be heard after the expiration of a period of 15 years. If they relate to real property dedicated to pious purposes, however, the trustees thereof are entitled to bring an action relating thereto up to a period of 36 years. Actions relating to the government land and actions relating to private roads, to a right of flow and to a right of taking water, if they concern government land, shall not be heard after the expiration of a period of 10 years since action was last taken in connection therewith.[167]

This code has ceased to be in existence owing to the fall of the Ottoman Empire, and some of the countries in which they were prevalent have come up with newer laws based on Shari'a.

---

[164] Article 1324, The Mejelle, available at http://www.iium.edu.my/deed/lawbase/al_majalle/index.html, last accessed on 03/06/2016.

[165] Article 1325, The Mejelle, available at http://www.iium.edu.my/deed/lawbase/al_majalle/index.html, last accessed on 03/06/2016.

[166] Article 1326, The Mejelle, available at http://www.iium.edu.my/deed/lawbase/al_majalle/index.html, last accessed on 03/06/2016.

[167] Article 1662, The Mejelle, available at http://www.iium.edu.my/deed/lawbase/al_majalle/index.html, last accessed on 03/06/2016.

While we go through the code, as such, we come to realise that water and its governance and use was subject to a lot of thought and the very presence of laws which regulated the use and laid down conditions and prescribed methods and ways in which it was to be enjoyed were quite comprehensive. All these acted as precursors to the law that we have at present besides playing a great role in evolving a comprehensive regime as far as water law and water governance are concerned.

Observations

Water, having been treated as a lifeline of civilisations over a vast period of time, has been sought to be protected and put to a judicious use so that it is preserved for future generations as well. Having come to know of the importance attached to it, a myriad of steps have been taken highlighting its importance and spreading the same across.

Right from community participation to various steps initiated as regards the same, water and water protection have been treated as a high priority. Instances of mobilisation of a group of farmers whose main aim is to formulate and implement community irrigation projects led to the birth of the concept of Pani Panchayats. The term was used for the first time to denote the five-member committee which was made to overlook the first lift irrigation project under the aegis of the Gram Gaurav Pratishthan (GGP).[168] From then, it has come a long way.

Considering the fact that India was a poor country and most of its population lived in villages, rural poverty alleviation was an issue that needed urgent attention. Providing for water for irrigation purposes soon took the forefront, and this led to the construction of dams, canals, tanks and the like that were needed to store water. To overcome the social indifference that was prevalent in India, the GGP laid down five main principles that can be traced to the working of the Pani Panchayats.

1. Irrigation schemes are to be taken up not by individuals but by groups. Water would be allocated as per the number of persons in the family and not as per the land holding.
2. To prevent the abuse of water, it has been laid down that only seasonal crops with low water requirements are cultivated.
3. Water rights stay with the farmers even if the land is sold.
4. Everyone, even those who do not own land, have a right over water.
5. Beneficiaries of the Panchayat have to bear 20% of the cost.[169]

[168] For more, see http://panipanchayat.org/content/gram-gaurav-pratishthan, last accessed on 04/06/2016.

[169] See http://www.goodnewsindia.com/Pages/content/inspirational/paniPanchayat.htm and http://www. rcdcindia.org/PbDocument/2ad288da9a427e8-2607-4a3d-93e9-71f47b208f6dPANI%20 PANCHAYAT%20REPO RT-%20FINAL%20(1).pdf, last accessed on 04/06/2016.

Based on such steps taken by the Panchayats, it was ensured that everyone had access to adequate water. It was also ensured that, since this was a people-oriented initiative and had maximum participation, in its true sense, the people themselves would work for the betterment of the whole community as they too gained from such an exercise.

It can be seen that such steps have gone a long way and have promoted the sustainable use of water. Steps for better water management can be found in the building of johads,[170] which has played a great role in raising the water level in arid regions. Since this method has gained prominence, more and more villages have come forward with the idea and tried to improve their quality of life by building johads and thereby being self-reliant.

It can be said that riparian rights based on common property resources, usufruct rights and controlled and reasonable use of resources have guided human settlements. Taking clue from these, it can be seen that in Northeast India, the old riparian rights system of dongs helped in controlling water use.[171] Similar methods like ahars and pynes were revived in Southern Bihar. The former was an artificial network of community field channels that helped divert water from rivers. It is through such pynes that rainwater was also being harvested. They served two purposes too. One, the bigger ones carried flood waters over long distances, and the smaller ones helped irrigate lands around a small area. Ahars on the other hand were man-made reservoirs that acted as community water harvesting tanks. It is quite interesting to note that such ahars were also created during the period of Chandragupta Maurya. Early riparian rights were based on the concept of sharing and conservation.

It can also be seen that in South India, community-managed water systems were present. Such community-managed water systems, called kudimaramath,[172] were aplenty. However, the sad part is that the same did not take off well and was not as successful as other community-based efforts. Due to the fact that it did not gain much attention and was incapable of harnessing adequate community participation, the system failed to rebuild the water commons and fell short of its objective.

Even though it is interesting to note that such a concept was prevalent in South India. Called Panchayats in Karnataka and Nattamai in Tamil Nadu, these systems were born out of farmers associations and helped a great deal in the construction, maintenance and upkeep of water systems and water resources.

In cases where water resources served more than one village or area, a representative body was formed to ensure that a democratic way was put forward for an equitable distribution and use of water resources. For example, in the Himalayas, where Kuhls were the main source of community water supply, a democratic use of the same was ensured by appointing irrigation managers. It is also interesting to note that such people were chosen from landless Harijans who were given full

---

[170] Water harvesting structures.

[171] See http://hindi.indiawaterportal.org/node/53132 , last accessed on 04/06/2016.

[172] The term literally meant 'self-repair'.

autonomy to open and close tanks. There were also specially appointed persons called *neergantis* who ensured that water was distributed and was made available equitably.

The main aspects that need be looked into as far as evolving an effective water management system are:

- Augmentation
- Protection and prevention of pollution
- Storage
- Distribution
- Conservation

Reformative punishment aims to make man understand the wrong in his acts and tries to create a sense of repentance in man when he realises his mistake. Law, as such, many a time, lacks this sort of an effect on man. The course of the study shows that law has not been able to stop man from destroying the environment, owing to his wanton acts.

A peek into the grim situation shows that there was no dearth of laws on the subject. Many a nation, sensing the significance and magnitude of the problem had made laws. India, too, had, owing to the emergence of the area of environmental law, enacted many laws which sought the protection of the environment, by regulating the acts of humans and prescribing punishment for certain acts. The world had grasped the gravity of the predicament that it posed.

Punishments for environmental degradation were galore, and they were made out for a variety of offences, under numerous laws. The enforcement mechanism, though not satisfactory in implementation, made its presence felt. Various bookings under various acts were made, and lots of people were punished. Environmental protection was the watchword. Thus being so, people were aware of the consequences they would face, if they did anything which was in contravention of the law present.

But was the so-called law successful in protecting the environment by deterring man from his anti-environment acts? The answer is a big and blatant, on the face, *no*! This is made evident by the innumerable circumstances wherein he was plundering the resources given by nature, shamelessly and consciously. Mother Nature was suffering at the very behest of her 'loving' sons and daughters! Law, with all the sanction that it prescribed, was not able to restrict man from harming the environment.

For sure, law made its presence felt when people were punished for actions harming the environment. But then still people continue doing so. This is because formal laws do not have the effect of creating awareness; as for the common man, it is made by the rulers and imposed on him. This situation happens even in a democratic society, as the common man's participation is very limited in law making. Law, thus, remains aloof from the people for whom it is made. The compliance is ensured not through a hearty cooperation of the people but through coercive measures. This is the greatest limitation of law in any case and particularly so in the case of protection of environment

But why has law failed? This calls for a thorough introspection. With the onset of laws, it can be seen that man was concerned more with his rights. Duties, a component of similar status, if not more, were completely overlooked by man. Overshadowed by rights, duties had been treated with 'contempt', if liberty can be taken to use such strong words. This 'alien' treatment of duties is what has been the main reason for the failure of law.

The reason for this failure can be attributed, as said earlier, to the compulsory nature of law, the 'imposition' of it on man. Law is an external agency. It is a child of the state. It has not come from within. Had it come from within, it would have definitely played a greater part in controlling man's actions.

But just for the fact that law has failed, it does not mean that we are not to seek refuge in some other factor; with environmental degradation posing an immediate threat, it is quite imperative that we look into other avenues wherein some succour can be afforded to. Where the present law has failed, what could possibly succeed? What could probably reinvigorate law and ensure its compliance?

An alternative for law was sought and the work dealt with religion being mooted as the source. Religion, as we all are aware of, has a control over its followers that is far beyond common man's perceptions. It has been able to mould man and restrict his activities in a way that is conducive to its teachings. Religion thereby exerts a much greater influence on man than laws at present.

Now the inevitable question arises. What is it that religion has that law lacks? It is this secret ingredient of religion that ensures that its principles are followed. Law, which provides man with the rights, it can be seen, fails miserably in the task of instilling duties in man. Why, even though the environment is at the receiving end of man's wanton acts, man still took such a lot of time to realise it is made clear when one understands that man was too busy availing of his rights that he forgot his duties.

Whatever be said and discussed, it is to be taken into consideration that law has to reinvent itself and change along with the changing society, societal concepts and notions. From the various teachings that religion provides, law must be able to imbibe the true spirit and try to protect those aspects that people keep close to their heart. Let religion be used as a tool to foster a better environmental protection and law harnessing such a benefit and gaining recognition and compliance rapidly.

What has to be achieved to gain success is to ensure that the legal approach towards environment should have a paradigm shift; a shift that would necessarily make it a 'duty'-oriented perspective than a 'right'-oriented perspective. Reaching such a high pedestal takes its own time and effort, but man has no other choice than to follow the high levels of compliance lest, mankind as a whole, suffer and perish. If such a shift is possible, then it can be said that the resources are in safe hands and would stand the test of time.

For making the same a reality, it is high time the society looked back into the practices that they had and the steps that were taken and try to bring about a revival of the same or reinvigorate the same so as to ensure a better protection of natural resources, in this case, *water—the lifeline*.

Acknowledgements The author would like to express his profound gratitude to Prof. (Dr.) M.K. Ramesh, NLSIU, for having guided and supported him, at every stage, and help him complete this work. The author also acknowledges the support rendered by Prof. (Dr.) Abdul Azeez, Prof. (Dr.) K.V. Raju, Dr. S. Manasi, the co-authors and all others engaged in this effort.

Water in Scriptures: A Comparative Overview

K.V. Raju, S. Manasi, Abdul Aziz, M.K. Ramesh, and M.S. Shruthi

Background

Water evokes a variety of emotions. It conjures to the mind's eye the images of conflict among states, chaos in governance, famine and floods besides conditions, concepts and ideas alike, basic need, entitlement, life support, commodity, etc. Concerns for conservation and a lurking danger of future wars becoming a reality over this scarce resource contribute to a disturbing kaleidoscope and collage of images of perception, use and governance.

Ecological and human rights concerns dominate international strategies and discourses on water management. While, at the global level, the need for an integrated water resource management is stressed, at the European regional level, a Framework Water Directive Model is presented, as an answer for addressing the problems associated with water governance. Back home, serious discussions are on at the highest level of decision-making, as to what shape, form and texture of water governance suits India best. The Planning Commission and the Ministry of Water Resources at the centre have, each, come up with a framework law for water governance in India. Some of the states are visualising a comprehensive water law (as is being proposed

K.V. Raju
Policy and Impact, Asia Regional Program, International Crops Research Institute for the Semi-Arid Tropics, Hyderabad, India

S. Manasi (✉)
Centre for Research in Urban Affairs, Institute for Social and Economic Change (ISEC), Bengaluru, India
e-mail: manasi@isec.ac.in

A. Aziz • M.K. Ramesh
National Law School of India University, Bengaluru, India

M.S. Shruthi
Research Assistant, Institute for Social and Economic Change (ISEC), Bengaluru, India

K.V. Raju, S. Manasi (eds.), *Water and Scriptures*,
DOI 10.1007/978-3-319-50562-6_6

for Rajasthan) and River Basin Management Regime (the blueprint of which is being designed for Meghalaya). Reflections like these, if not anything else, bring to a sharp focus the crisis in water governance and its growing scarcity. This also makes one look for inspirations and better models of governance—internally, internationally and comparatively—to adapt and emulate. The quest for an alternative, viable and sustainable model of management that would ensure equity in access to and use of water dominates such an effort.

Given this backdrop, there is a rising interest in values and ethics related to policy that cuts across academic disciplines and stakeholders. The need has been to widen the focus with culture and religion playing an important role in conserving and protecting environment. Towards this, there are several initiatives taken across the world trying to understand the relations and interlinkages between culture and development. An interesting study since 1998 (Bertus and Reijntjses 2007), between the partners from Latin America, Africa, Asia and Europe, has been on intercultural dialogues on developmental approaches focusing on the south. They have studied the influence of world views on material and sociocultural as also technological development, scientific approaches and policies across different cultural contexts. The study has revealed that in most non-Western world views, there is a notion of three interrelated worlds—the material, the social and the spiritual. Based on these, it is emphasised that sustainable development is possible only with a balance between these three aspects—the material, social and the spiritual—and have called this approach 'Endogenous Development'. They emphasise that there is a need for encouraging the use of models available in our cultural religious scriptures which are worth a revisit and applied to the contemporary context. Shifting world views and values, new thinking and holistic theories, intercultural, inter-scientific and interreligious dialogues and methods of social learnings are considered most important in this regard. The need to widen the focus of science is demonstrated by a critical point in the evolution of the globe: the social, economic, cultural and spiritual crises have their roots in human behaviour, values and the use of knowledge and technologies.

With reference to water, there are several initiatives as well. The *Water Ethics Network* of the Water and Culture Institute seeks to stimulate debates and collaborative actions among the growing set of interested parties. The aim of the network is to bring the study of water ethics into the everyday discourse of water policies and management decisions, to ensure that the choices about water use and water ecosystem management are consciously informed by values.

Box 1: Water and Culture Messages

On the occasion of World Water Day 2006

1. Promote the understanding that water and culture are inseparable elements of human life. Culture should be regarded as the permanently evolving set of distinctive spiritual, material, intellectual and emotional features of society or a social group. It encompasses—in addition to art and litera-

ture—lifestyles, ways of living together, value systems, traditions and beliefs (UNESCO Universal Declaration on Cultural Diversity, UNESCO, Paris 2002).
2. Foster the dialogue of cultures to find solutions for water-related problems. Cultural diversity, stakeholder involvement and intercultural dialogue should be the guiding principles for the development of awareness raising, educational and capacity building materials and methods.
3. Promote inclusive, solution-oriented water governance that takes into account all facets of cultural diversity and that seeks informed consensus. Indigenous knowledge holders should be involved as full partners.
4. Encourage the artistic expression on water issues as an important means of fostering understanding and sharing information. Drawings, photographs, audiovisual materials and the performing arts often help to get messages across cultural and language borders.
5. Encourage the creation, transition and dissemination of information on water and culture in your community. Involve schools, universities, those in your work place and at home.

Source: UNESCO 2006.

The other initiative has been the development of *Water Ethics Charter*. The purpose of Water Ethics Charter is to provide a moral basis for water management decisions which cannot be accurately valued in financial terms and are not mandated legally. This is developed in partnership with the Academie de I'Eau, UNESCO and other organisations and in consultation with key water stakeholders (companies, cities, government agencies, indigenous people, NGOs, etc.). A consensus will be developed around a clear set of ethical principles for governing water use and management. The Charter will be presented at the World Water Forum in South Korea in April 2015 followed by a campaign to build awareness regarding the Charter besides inviting endorsements. Water Ethics Charter will serve as a tool for better decision-making through crystallising ethical concepts that make intuitive sense to a broad range of stakeholders, cutting across cultural boundaries and gender, age and class distinctions. The Charter will articulate a common set of principles about how water and water ecosystems should be utilised and protected for the benefit of present and future generations (Water Culture Institute 2013).

Box 2: Objectives of Water Ethics Charter

- Articulate clear principles and guidelines of what constitutes ethical water practices in particular situations.
- Recognise inherent rights of water ecosystems to exist in a healthy state and the right of people to enjoy clean water and healthy water ecosystems.

- Raise awareness about the ethical implications of water policies and water decisions at multiple scales.
- Elicit endorsement by companies, agencies, cities, indigenous groups and NGOs to uphold and disseminate the Charter.
- Promote social and legal reconciliation among diverse stakeholders through values-based deliberation and consensus around a shared water ethic.

Source: Water Culture Institute, 2013.

All these signify the emphasis laid on water and culture and its importance for improving and ensuring sustainable water resources management. Besides, it would be important and interesting to understand how water has evolved into involving various disciplines.

Historical Overview

In the context of water, a historical overview of involvement of various disciplines and their focus will provide us the need and importance of understanding water in its multiple contexts. In the early part of the twentieth century, engineers played the lead role, in that they designed and built dams, irrigation schemes, water supply systems and flood control works. The issues they addressed were primarily technical, identifying the best possible location for a given dam or the contemplation about the size of the channel or best ways to supply water to irrigation schemes. This was followed by the involvement of 'economists' in the 1920s and 1930s wherein projects were required to be justified on economic grounds. The question of cost-benefit of water projects came into fore, and the issues became financial wherein they had to justify the expenditure on a particular dam to be constructed, most economic route for the proposed channel, the period of expenditure recovery of a particular scheme and so on. Thus, 'economists' came to be more involved, while the 'engineers' embraced the cost-benefit analysis (Manthrithilake 2012).

In the 1960s and 1970s, 'environmentalists' came to the forefront with a major focus on the environment, particularly with anthropological impacts. Environmental protection agencies mushroomed in many countries. A number of water-related projects were slumped off due to environmental reasons. Thus, water resources planning became explicitly multiobjective, with engineers and planners being required to consider environmental and social objectives in addition to economic efficiency. Many other experts—ecologists, biologists, chemists, geomorphologists and sociologists got involved in water resources development. However, engineers continued to assume the leading role and often coordinated other experts in achieving the overall objective.

With 'sustainable management' becoming the main focus, the really critical resources were the nonrenewable ones. For instance, if a biological species is made extinct, it is impossible to recreate it; similarly, fossil fuels can be replaced at an infinitesimally small rate. Although fresh water is limited in quantity, it is infinitely renewable. This suggests that the sustainability of water resources is secured when compared with biodiversity and fossil fuels. Following these discussions, integrated water resources management came to the forefront. There were discussions on the conflict between water use and food production. To achieve sustainable use of water resources in the future, the world's population had to be stabilised at a sustainable level. There was need for a greater emphasis on water reuse and recycling. There was likely to be a greater integration of the planning of water, land, and biological resources at the basin level in the twenty-first century. This can be seen as a natural progression of integrated catchment management and is a promising trend (Manthrithilake 2012).

By the 1970s–1980s, engineers were no longer allowed to be taken for granted that they knew what was best for the community. Consultations with communities became very important with participation becoming the mantra. It was recognised that value judgements were used in selecting projects. Stakeholder consultations were to be done in a systematic way to determine community preferences. This required inputs from new kinds of expertise to conducting consultations wherein lawyers and gender experts were to be involved in decision-making. As more than 50% of the world's population is women, it is seen that they have different value sets than men (as water consumers). Women place a high value on human life and value the environment very highly. Later, we had the diplomats who were to be water resources managers for handling the concerns with regard to water diplomatically, given the interstate and intercountry sharing of waters. Spiritual and aesthetic values of water have been very pertinent in the ancient scriptures with frequent references made to in the daily routine rituals being practised by communities in the hoary past. The spiritual and healing effects of water have re-evolved into holistic healing methods adopted by several healers in their therapies. We see the role of artists in water resources management with several movies made on water to create awareness and values among people towards water. Theologians have also involved themselves in the study of water. Musicians have always played a major role in water; Jalatarang is an instrument based in water, so also, musicians singing songs related to water has been quite prominent. We may have more disciplines get involved in water in the coming future. Literature across languages, be it English or local languages, has poems/verses/songs written on water highlighting the significance and relevance of water in various contexts.

The status of health and poverty eradication has cultural nuances; culture has positive and negative health impacts on individual well-being, more so with women's health. Water management practices should be adapted to specific cultures as they constitute distinct systems of knowledge and behaviour. Hence, water resources management strategies must take culture fully into account.

With culture and varied disciplines getting into the forefront of water management, it is important that we look at the relationship between water, culture and religion and the specificities they have to offer towards water management and the commonalities across the three religions that have been studied so far.

Water, Culture and Religion

Water has a significant role in the lives of people with many religions of the world recognising this significance associated with it (Stewart 2008). Interestingly, water is considered important not only during the life of a person but also after one's demise (Oestigaard 2009). Hence, water is part of various rituals performed during birth and death of individuals. Water signifies and symbolises purification or cleansing of sins. For instance, the ritual baptism, a person's entry into the church, is revolved around the water wherein the person is immersed in water or a few drops of water sprinkled over the person symbolising purification or removal of sins. Water is also considered the spirit of God in the New Testament of the Bible. Similarly, in Islam, water is used for purification and performance of three types of ablutions—washing of the whole body, head, hands, forearms and feet before prayers. Hindus also place a strong emphasis on purification in that they believe water purifies one's spirit and soul and that it is important for people to bathe before entering into a temple.

Scriptures and Sources: Islam, Christian and Hindu Religions

Origin and sources of these most popular and practiced religions in the world were evolved centuries ago. These religions have religious texts/scriptures that are written and specified doctrines/principles/guidelines that are to be followed. Besides conveying various other moral and ethical values to society, it is important that these ancient scriptures across the religions have made several references to natural resources management which highlights their sustainable management. The following section provides an overview about the origin of the religion, holy books and other associated scriptures.

Islam

Islamic philosophy is the product of a complex intellectual process of Syrians, Arabs, Persians, Turks, Berbers and others. Traced chiefly to the Arabian Peninsula, an arid desert part of the planet earth, the pivot around which the whole of Islamic life turns, Islamic philosophy is around the Quran, revealed to Prophet Muhammad by God through Angel Jibreel between 610 CE and 632 CE.

The Quran is supplemented, however, by a mass of utterances attributed to Muhammad that constitute together with circumstantial reports of the actions and decisions of the prophet, the general body of Muhammadan Traditions, properly designated in Muslim usage as the prophetic 'way' (*al-Sunnah*). The sayings of the prophet come next in the hierarchy called as Hadith (pronounced as Hadees) compiled 200 years after the death of prophet. Fiqh is another scriptural authority, of which a huge collection is produced, employed to extract the jurisprudent practices in the early Islamic societies.

Christianity

The word Christianity is derived from the Ancient Greek word Χριστός, Christos, a translation of the Hebrew Mashiach meaning 'the anointed one', together with the Latin suffixes -ian and -itas is a monotheistic religion based on the life and oral teachings of Jesus. A predecessor of Islam traced from the first century CE with the birth of Jesus Christ. Christianity is an Abrahamic religion[1] that began as a Jewish sect in the mid-first century originating in the Levant[2] region of the Middle East.

Christianity regards the biblical canon; the Biblical writings are traced to from the eighth century Before Common Era or BCE. The Bible is called as a book of many books segregated into Old and New testaments. The Old Testament consists of 39 books, constitutes the holy scriptures of Jews and focuses on human history. The New Testament concentrating on the teachings of Jesus Christ contains 27 books.

Hinduism

Hinduism or precisely the Vedic thought has guided man from time immemorial; many practitioners refer to Hinduism as 'the eternal law' or the 'eternal way' beyond human origins. Many scholars regard Hinduism as a fusion or synthesis of various cultures and traditions with diverse roots and no single founder traced primarily from the Indian subcontinent.

'The term 'Hindu' first occurs as a Persian geographical term for the people who lived beyond the Indus River (Sindhu in Sanskrit)'. Hence, the term 'Hindu' is a

[1]Abrahamic religions (also Abrahamism) are the monotheistic faiths of Middle Eastern origin, emphasising and tracing their common origin to Abraham (Abraham is a key figure in the religious texts of Judaism and Christianity as well as Islam's holy book the Quran) or recognising a spiritual tradition identified with him. They are one of the major divisions in comparative religion, along with Indian religions (Dharmic) and East Asian Taoist religions.

[2]The Levant, also known as the Eastern Mediterranean, is a geographic and cultural region consisting of the 'eastern Mediterranean littoral between Anatolia and Egypt'. The Levant today consists of Cyprus, Lebanon, Syria, Palestine, Jordan, Israel and part of southern Turkey.

geographical term and does not refer to a religion. Hinduism is fundamentally based on 'the accumulated treasury of spiritual laws discovered by different persons in different times' called Veda which forms the earliest and the most authentic record of the Hindu scriptures. The Veda is broadly divided into four chapters called the Rigveda samhita, Yajurveda samhita, Samaveda samhita and Atharvaveda samhita. All these four chapters are developed with a three-tier ensemble of commentaries, usually in prose, and are believed to be slightly later in age, namely, Brahmanas, Aranyakas and Upanishads. These are collectively called as 'Shruti' (that which is heard).

The next in the chain of command of Hindu texts are called the Smritis (memory) which deal with the ceremonial doctrines to be observed written in accordance with time periods. The most notable of the Smritis are the Manusmrithi, Parasharasmrithi and Narada Smrithi. Apart from Shruthi and Smrithi, epics which consist of the Rāmāyana, Mahābhārata and the Bhagavad Gītā, an integral part of the Mahabharata, are also the most popular sacred texts of Hinduism. The scientific literature developed on various subjects such as health, architecture, political economy, chemistry, literature, engineering and astronomy constitute the stream of Hindu scriptures in the lower hierarchy.

Veda postulates God is only one for the whole universe, all powerful, all knowing and omnipresent and works completely selflessly to facilitate the individual entities or souls in attaining 'Moksha', a considerably eternal state of bliss which could be achieved following the righteous path shown by the supreme power in one's life. Hence, everything in the universe is a manifestation of 'God'. As such a Hindu worships everything, from nature to art to tangible objects such as idols.

All the sciences developed by Hindu civilisation are the derivatives of Veda. Both physical and cosmic sciences developed are said to be about 64 in number and only a few of them have the literature available as of today. Hinduism is a faith followed from as long as 10,000 years, and it was not started by one person at one point of time. Hindus find god manifesting himself through this creation and hence find him in everything in the nature and worship everything.

Water in Scriptures

All the three scriptures, viz. Bible, Quran and Hindu scriptures like Vedas, Upanishads, Bhagavad Gitsa and Puranas do talk about water in their various ayaths, slokas or messages. In the first phase, as far as the origin of water is concerned, Quran, Bible and Hindu scriptures suggest that, on our earth, water originated first and was then followed by plants, trees, animals and humans. In other words, according to the scriptures, life emerged only after water was created. Interestingly Bible and Vedas suggest that life emerged from out of water which actually is consistent with what our scientists also suggest. This means, water is the life-giving and also the life-sustaining source—a proposition that cannot be disputed. For this reason, Bible treats water as sacred. Being so it regards water as a blessing from God in so far as it is believed to have come from heaven. Also Bible picturises water as a rescuing force as when an exodus took place under the leadership of Moses. The opposite that it could be a force of destruction is also true as one could recall the anecdote of deluge during the age of Noah.

Besides, all the scriptures have seen water as a divine element, a sacred gift of God. Hence, water is recognised as a spiritual symbol by all and worshipped, particularly by the Hindus. They also speak of the benefits derived in terms of fertility, water flow and so on. However, there exist some differences in their perceptions regarding 'Water is God' or 'created by God'. The Vedas have also described about the origin of water; water is said to have emerged out of fire that came after gases which came from ether that emerged out of 'Ambhah', the most primordial matter that was and will be present in the course of creation in various transformations. Importance of water is depicted in a variety of ways. There is a scintillating account of Science and Omniscience of Water. Prayers are described as expressions of 'gratitude towards water, and literary works are cited to drive home the point that water was the first of creations. The section on the 'qualities of water' in Hindu scriptures is worthy of a special mention. There is this insightful reflection on 'Chaturvarna' (the four occupational categories or castes), that is, both novel and rational. The authors assert that the 'Vaisysa', among the four Vernas, had the obligations of maintenance, conservation, value enhancement, improvement of quality and judicious distribution of resources. While, this was primarily the task of one Verna, it was a common obligation imposed on every one not to pollute, defile, degrade or in any way affect the quality of water bodies since 'water was precious and belonged to all'.

As stated (Table 1), since life emerged from water, the basic quality of water is that it not only gives rise to life but also sustains it. The Rig Veda says that water is a compound but does not explain what this compound consists of. Scientists observe that water is a compound consisting of two units of hydrogen and one unit of oxygen. Apart from the life-giving quality, water has other qualities to which some reference is available in both the Hindu and Christian scriptures. Both these scriptures recognise the potential of water as medicine and also as a compound that cures diseases. Thus, Ayurveda recognises water as a compound that cures many diseases and hence suggests a full bath of the body for curing of its external ailments. Even with regard to treating internal ailments, Ayurveda has natural medicines particularly prepared from medicinal plants. However, in the preparation of medicines, water becomes an important component in so far as it either comes as a juice in the lehya or is added to the lehya. There are medicines that target ailments arising on account of vatha, pittha and vayu. For the so-called hyperacidic ailment, there are of course ayurvedic medicines, but Ayurveda practitioners also suggest water therapy, i.e. drinking large quantities on empty stomach in the morning and removing the same forcibly so that the excess acid is brought out. Similarly, in respect of respiratory problems, water therapy is recommended, i.e. taking in water through one nostril and bringing it out through another which clears the nasal canal, making way for an easy entry and exit of air to the lungs.

Bible regards water as a healing and cleansing medium/agent of ailments when it cites the examples of how the sick were healed by dipping themselves in the

Table 1 Perceptions regarding water

Islamic	Christian	Hindu
Water a divine element		
Creation of God	• Created by God	Water is God
	• Regarded as a blessing	Water was made during the time of creation
	• Jesus performed first miracle using water	
	• Spiritual symbol outpouring God's spirit	
	• God causes rains and is compared with rain	
Creation of life		
Cause of primary creation (Quaran-24.45, 25.54)	Gave birth to all life forms	Gave birth to all sources of life
Qualities of water		
• Water sustains life (Quaran-15.22, 16.10, 16.11, 43.11)	• Source of life and rescuer	• Described as 'Jeevanam' meaning life
• Water, a food of plants, a fertiliser (15.22)	• Thirst quencher	• Forms the main constituent of majority the things in the universe
	• Water preexists in all forms of nature	• Everything in the universe is formed by a combination of five elements—ether, air, fire, water and earth— and have different qualities
	• Living water (river, spring)—in Judiac law	• Around 100 words synonyms for water in Sanskrit and Vedas describing qualities of water
	• Jesus is considered as the source of living water (John 4.14)	• Hindu philosophy describes water as characterised by taste and sensed by tongue
	• Water cleanser and healer (Leviticus 14:8-9)	
Referred metaphorically and symbolically		
Symbolises paradise	Used literally and figuratively	Figuratively used in various scriptures and writings
Perceived benefits of water		
• A road for navigation (45.12, 35.12)	Not available	• Described as utter necessity for consumption, cleansing, relieving, hospitality, farming and consoling
• Irrigation		• Maritime trade
• Used to improve aesthetics		• Used for music composition—Jaltarang
		• Water clock
Medicinal properties of water—beliefs and practices		
Water of the Zamzam spring in Makkah has curative properties	Water is considered a healer	Universal medicine

Jordan River. Also Jesus healed a person afflicted by leprosy by sprinkling water on him, and also he sent a blind man to a pool to wash himself up whereupon he was said to have regained his eyesight. In the Islamic societies too, water's healing quality is recognised in so far as Hakims and unani doctors use water as an ingredient in their medicinal preparations.

In addition, since water is seen in Puranas as an avatar of Ganga, Ganga water is considered sacred; people believe that a dip in Ganga washes off one's sins. This belief also is behind the practice of washing dead bodies with Ganga water and bringing water home all the way from Ganga for performing various rituals. Among the Christians, water is believed to be so powerful that Jesus Christ cured leprosy and other skin-related diseases by sprinkling water on patients. Among Muslims, there is no such belief in the power of water, but they consider water an important cleansing agent of the human body and surroundings of the residential areas. However, those of the Muslims who go to Mecca to perform Haj (pilgrimage) bring water from a spring which is described as Aab-e-ZamZam and which is regarded as holy water. But by and large, Muslims use water for the ritual of cleansing oneself, particularly just before starting the daily obligatory five prayers. Also it is obligatory on the part of the Muslims to have a ritual water bath every time he/she has had sexual intercourse. In this context, Prophet Mohammed had suggested that Muslims should use clean water for ritual bath and ablution. According to him, running water such as rainwater or river water is clean. While stagnant water is not. However, water stored in Hauzes (pools) in mosques for ablution is not treated as unclean provided Hauzes have a prescribed dimension. In the Hindu scriptures too, temples do have water storage/pools called pushkarnis where water is available for cleansing and bathing purpose. In fact, this water is also used for bathing idols and washing the surroundings, an indication to the effect that pushkarni water is clean and sacred.

Among Christians too, water is considered sacred and a cleansing and purifying element. Thus, in the Christian community, priests are washed at their consecration ceremonies; special ablutions are expected of the Chief Priest on the Day of Atonement; Baptisation of persons involves either immersion in deep water or sprinkling of water. Thus, such rituals highlight the importance of water among the Christians.

Apart from the religious significance of water, all the three societies—Hindu, Muslim and Christian—recognise as stated above the life-giving quality of water. As a follow-up to this, in all the societies, water has been used for irrigation purpose for cultivating crops and growing trees and other plants. Besides, water is consumed by animals, birds, humans and other living organisms for sustaining their life. Since water is consumed for sustaining life, the religious scriptures prescribe that purity and cleanliness of water be maintained. Prophet Mohammed is quoted to have decreed that no one urinated in water, emptied bowels near water sources and bathe in stagnant pools. Human civilisations with a view to ensuring pure water for human consumption have, at different stages, evolved and followed different practices of water purification. These include guarding water against organic waste and other debris falling into it dropping alum into water to hasten the process of particulate matter to settle down or sand filtering of water.

Qualities of water are described in its various forms across the scriptures of all the three religions. While Islamic scriptures indicate the physical uses of water for plant growth, the Christian scriptures highlight water as a source of life and flowing water as living water, thus not encouraging the use of stagnant water. Water is also seen as a cleanser and a healer. In the Hindu scriptures, consuming water is seen as providing freshness and enthusiasm. Besides, water is related to one of the forms of earth's creation, and there are several synonyms implying the qualities of water.

All the scriptures are replete with references about water. As can be seen, all the scriptures have largely common references to water and accept water as the primary source of life creation and sustenance and hence the primary raw material necessary for raising food for life, man, animals and plants.

Water as a healing element is seen prominently in Hindu scriptures compared to Islamic and Biblical scriptures. There is a separate subscript 'Ayurveda' which makes references to the various uses and types of water in different seasons. Various types of water include hot, cold, heated and cooled and rainwater with their qualities, and water at different seasons is referred to in the texts of Ayurveda and is used both as base medicine and independent medicine. In Islamic literature, there are references to spring water as holy water having medicinal properties, while water is considered a healer in Biblical scriptures. Thus, importance of water as a healer is found across all the scriptures.

Prayers on Water and Its Use of Water in Social and Religious Traditions

Since water is considered by Hindu scriptures as sacred, members of the Hindu community do pray for water. Every literature produced by ancient Hindus is in the form of prayers. Prayer not only indicates submitting oneself to the almighty but also committing oneself to do what is required to achieve a need. Water, a very beautiful and precious part of nature, is revered with utmost devotion. There are several works dedicated to glorifying, conserving and protecting water too. Thus, they offer prayers to water two times a day, first in the morning before sunrise and second in the evening before sunset, and are called prathas sandyavandanam and saayam sandyavandanam, respectively. Also both when the body is cleansed and idols washed, water is used by chanting prayers called sookthas. On the other hand, in Islam, a monotheistic religion, it is not permitted to pray to anything that is created including water. Islam permits prayer to God alone, not his creation. Hence, prayer in respect of water is prohibited. However, water is used as a medium of cleansing one's own body by performing vadhu (ablution) and ghusul (ritualistic bath). Although water is thus recognised as an important medium for ritualistic purposes, prayers for it are not permitted by Islam. However, praying for rains is referred to in Islamic literature as also Hindu literature through yaagas and yagnas to invoke rains. The Christian religion, which precedes Islam and in a way shares many commonalities with Islam, also does not seem to suggest any kind of prayer in respect of water, though Jesus Christ used water as a medium to cure diseases such as leprosy by sprinkling it on patients.

Across all the three religions, water is used while conducting various rituals and rites. Sacred ablutions were/are part of every religion. In Hindu rituals, water is/was used on several occasions as mentioned above indicating the importance placed on water. Ceremonial features involve immersion or sprinkling of water besides symbolising purification, common across Christian and Hindu religions. Flowing water is considered as living water in Christian scriptures and hence considered pure and recommended for rituals (Table 2).

Table 2 Prayers on water and rituals

Islamic	Christian	Hindu
Prayers on water		
No prayers about water in scriptures	Not available	Innumerable hymns in Vedas and other literary works as a way of appreciating the usefulness of water
		Hymns have given instructions to protect and conserve it
Praying for rains		
Praying for rains is in practice (namaz e istisqaa)	Not available	Yaagas and yajnas performed for invoking rains and attain peace using water
Prophet Moses also prayed for water that gave birth to 12 springs (2.60)		
Dogmas and practices—during religious rites		
Ablution was/is mandatory before Namaaz which leading to the provision of cisterns/overhead tanks and taps in mosques	Ablutions are compulsory during baptism	Hindus being the worshippers of nature, use water in almost all religious rites:
		• Sacred ablutions
		• Strengthening of foetus
		• Tonsuring of birth hair
		• Taking a student to teacher
		• Graduation of a student
		• Wedding ceremony
		• Charity or act of giving (Daana)
Ceremonial feature		
	Involves	Involves
	• Immersion in deep water or	• Immersion in deep water
	• Sprinkling of water	• Sprinkling of water
		• Pouring water over the object of charity indicating the act of giving
Symbolises		
	Symbolises conversion of a person into a new person who decides to bury his sins and lead a new life under the guidance of God	• Purification
		• Cleansing of sins

Table 3 Mythological anecdotes

Islamic	Christian	Hindu
Anecdotes and its significance		
Zamzam is a holy spring in Makkah that gushed out when infant Ismail kicked earth crying due to thirst	Mythological anecdotes where water is used to depict conflicts between good and evil and God and Satan	Every river and many other sources of water are associated with anecdotes
	Believed that the world was placed between water above and water below	Stories associated with River Ganga and River Cauvery are most popular
Holy waters and practices		
Muslims going on Haj pilgrimage compulsorily drink this water and carry it back home	Pilgrims visiting holy places carry holy water back home	Pilgrims visiting holy places carry water back home
Believed to have even curative properties		Most popular is holy water from River Ganga stored by almost every Hindu at home
Water festivals		
Phool waalon ki sair started by Mumtaz Begum, the queen of Akbar Shah between 1808 and 1837	Sukkot is a Jewish water festival	• Ganga Dashahara
		• Kumbh Mela
		• Hariyali Amavasya
		• Tulasankramana
		• Hartalika Teej
		• Ganga Mahotsav
		• Ganga Dashami
		• Minjar Mela
		• Boat festival
		• Adiperukku
		• Chhath
		• Teppotsavam
Celebrated after rainy season	Celebrated annually for 7 days based on the sayings in Isaiah 12.3	Season wise, different times annually/one time as well

Fictions are enchanting and created to captivate the people into following certain practices (Table 3). Many a time, these anecdotes are associated with the usage of certain waters with medicinal properties. With the anecdotes associated with them, people strongly believed not only in consuming the water but also in protecting the source. In the Christian literature, water is used to depict conflict between good and evil which is interesting. In Hindu literature, there are innumerable anecdotes referring to rivers, rituals surrounding lakes, etc. Interestingly, across all the three religious scriptures, holy waters are carried back home during pilgrim visits, consumed and distributed to all.

Mythological Anecdotes

As stated previously, according to Hindu Puranas, water was brought from Heaven by Bhagiratha, by way of praying to Maheshwara with a view to wash off the ashes of Sagara, family members who had been burnt by the curse of sage Doorvasa. The idea of washing off their ashes was to ensure sadgathi (salvation) to their souls, which means that Ganga water was and is considered to possess a sin-washing quality and hence was and is regarded as sacred. There are similar legends associated with the other rivers of India, particularly, Sindhu, Saraswathi, Yemuna, Godavari, Narmada, Krishna, Bheema, Palgu, Sarayu, Sri Gandaki, Gomathi, Cauvery, Kapila, Prayaga and Nethravathi, of which Saraswathi is considered invisible.

Here, it may be of interest to state the case of how prayer to water yielded water for use. The tradition has it that in a small village called Kaivara of Kolar district in Karnataka, a farmer dug a well with the hope that he would have water for drinking, bathing and irrigation purpose. But unfortunately, even after going to a depth of the ground, he could not strike water. At that point of time, a saint known as Kaivara Thata, also known as Kaivara Narayana Guru, instantaneously composed a prayer on water and sang it himself, urging mother Ganga to show herself up. It is stated that at the conclusion of the song, water sprang up with a gush filling up the well. This suggests how if sacred souls pray to water the Goddess of water shows herself up.

The Bible also regards rivers of the Middle East as sacred. In the Old Testament, there is a reference to Nile River of Egypt, giving way to Moses and his followers to undertake the exodus. In the Islamic scriptures, there is a reference to how Ibrahim's wife Hagar, after giving birth to Ismail in a forest, searched for water, but nowhere was it to be found. Following her prayer to God, the new born infant playfully kicked its feet on the ground which, as a miracle, gave rise to a water spring. This water, popularly known as Aab-e-Zam Zam, is considered sacred by Muslims even today.

The above anecdotes talk of the positive aspects of water. In the Christian and Muslim religious scriptures, there is an interesting anecdote which also brings out the fury of water and its negative role. The reference here is to a deluge which was sent by God to destroy humans and animals who had gone away from God's mandate. The deluge took on the form of a huge wave, submerging everything that lived on earth.[3] However, Prophet Noah had saved some species of life such as humans, animals, plants, etc., by putting them on a boat constructed in anticipation of the deluge. Finally, the boat was carried by water up on to a mountain where it got stuck at a place where water would not reach. And that is how Noah is supposed to have saved life on earth from a complete destruction.

In Hindu religion, water festivals were, indeed, celebrations of joys of life, stages of life and the life after! Water festivals are followed across the length and breadth of the country with an emphasis is on water management techniques, conservation

[3] One may recall here the destruction recently caused by a tsunami.

techniques and discussions along with enjoying the blessings bestowed by water. However in Islam and Christianity, there are references made to one festival performed even to this day. As part of creating awareness regarding the importance of water, water festivals are held in some parts of the world including India. Coming to India, for example, at the confluence of Ganga and Yemuna at Allabahad, a very big water festival called 'Kumbha Mela' is held once in 12 years. Down south, in respect of Cauvery, at Bhagamandala where Cauvery emerges, water festival is held every year. In both the places, a large number of people congregate and celebrate the event in the form of a festival. On a smaller scale, water festivals are held at dam sites and irrigation tanks whenever these water bodies are filled up during rainy season. It is customary on such occasions for people in power and others to perform poojas and offer baginas respectfully, thanking goddess water. In the villages in South India where irrigation tanks are found, whenever these tanks are filled up following rains, there is a practice of holding what is called Teppothsava, the festival of boating. It is a beautiful spectacle to watch on such occasions how the entire village folk gather fully dressed in a festival mood to celebrate the occasion. Occasionally, animals are also sacrificed to propitiate the Goddess of water.

As stated previously, water is considered to be a cleansing agent and as such it is regarded as sacred. By reason of this, rivers are regarded as sacred and as such are respected. This quality of rivers is derived from mythological descriptions, especially the Hindu scriptures. Also, since rivers and water are respected, there have emerged dogmas and practices with regard to the use of water under social and religious traditions. As referred to above, water is used dogmatically in all the social and religious functions. And as such, water use has been in vogue all along the stages of human civilisation. Thus, when a child's birth takes place and when a child is given Brahmopadesha at the age of 8 years, when the child grows up to be an adult and enters into Gruhasthashram, when he or she attains the age of 60 when Shasthabdi is celebrated and finally when he/she leaves this world, water is used as a medium of cleansing the body by taking water as Theertha. Among Christians and Muslims, water is seen as a purifying agent at different stages of one's own life like when one is born and also when one is dead; in respect of the latter, it may be noted that dead bodies are ritualistically cleaned by water and this ritual is quite elaborate.

Promoting Conservation and Protection

Sustainable Practices

There are references in all the religious scriptures relating to the need for augmenting, storing, protecting, conserving and equitably distributing water (Table 4). As per the references made in the Hindu religious scriptures, Natural resources management was very simple and decentralised in the past. Several sensible measures used to be taken up voluntarily, such as religious and wisely through temple forests, maintenance of grazing lands, village tanks, de-silting of tanks, sustainable extraction of

Table 4 Sustainable practices for water conservation

Islamic	Christian	Hindu
• Islam always condemns wastage of water	• In Palestine, people maintained wells and springs	• Veda rules everyone should perform Devayajna (conserving natural resources) every day
	• Water was stored in cisterns (Exodus 21.33 and 34), public pools and aqueducts (2 Sam 2.13)	• Vyshya is specially assigned the job of conserving, enriching and distributing the natural resources
	• Most homes in Jerusalem had private cisterns (2 kings 18.31, proverbs 5.15)	• Natural Resource Management was simple and decentralised in the past; several measures were taken voluntarily. Common sources of water such as community wells, temple forests, grazing lands, waste water soaking pits and village tanks were constructed and maintained by ordinary citizens
		• Sustainable extraction of groundwater and village lakes was extensively promoted
		• Natural farming was practiced earlier which demanded minimal use of all resources including water
• Taking bath in stagnant water is forbidden	• Bible proscribes use of stagnant water	• A person should not take bath more than once a day under normal conditions
• Urinating/clearing bowels in water or near a waterbody is prohibited		• Water should never be denied to anybody
		• In the monsoon month of 'ashada' (July and August), conducting of religious ceremonies or celebrations is forbidden
		• Making drawings and designs of water on floor and walls is prohibited

groundwater, soaking pits, using simple and effective technologies. There is a detailed account of sustainable agricultural practices (natural farming) whose fertiliser and water requirements, in comparison with other practices, were the least. Community wells, as against individual bore wells, ensured a sustainable extraction of water for all. Islam is a proponent of water conservation, a scarce commodity in Saudi Arabia, a place where Islam took birth. Similarly, the Christian era also refers to several sustainable practices of water management. In the Christian and Islamic worlds, there has always been an emphasis on seeking clean water not only for consumption and cooking but also for bathing purposes. Religious scriptures of Islam and Christianity recognise the fact that rainwater is the cleanest of all the water sources. But the problem was that rains were seasonal besides being subject to a limited period of time. However, water was required throughout the year. Hence, people thought of storing water to meet their needs all through the year. The need for water storage arose in some regions for another reason which is the scarcity of water due to irregular or scanty rains. This was the case particularly in the Middle East where there was a vast desert with no abundant water resources. For these two reasons, the need for storage of water arose. Three categories of storage were in practice in the Middle East. One was storage of water in personal containers at the household level. Second, water sources are located on the privately owned properties such as ponds, tanks, spring, etc. and third on rivers, lakes and aquifers. In respect of the first two, the right of ownership of these sources rested with private persons but in respect of the third category of sources being the commons, the ownership rested with the community as a whole.

At this point, it may be of interest to refer to awareness levels among the ancient people regarding water cycle and hydrology. Bible holds that water for human and animal use is not available in the stock form but is provided by nature in the form of a flow. Thus, it suggests that rain and snow come down from heaven, but stay on earth for meeting the water needs of animals and plants. The excess water takes on the form of streams, flowing into the sea. And the seawater, in turn, returns to clouds such that clouds once again pour water on the earth. That is how the water cycle was understood by our ancestors.

Since water was considered to be a life-giving source and hence essential for life in itself, the scriptures prescribed that water should not be polluted either by urinating or excreting or even discharging pollutants. Also it was mandated that water should not be wasted. Thus in Islam, there is a concept of Khalifa, i.e. humans are said to have been created to take on the role of vicegerent of God on earth with the responsibility of promoting, preserving, conserving, managing and distributing the natural resources. In other words, man is considered as a trustee or a custodian of the natural resources in the sense that the use and distribution of water ought to be equitable. As a follow-up to this, Islam prohibits its followers from wasteful or overconsumption of water and even hoarding it to deprive its use by others. Therefore, it is mandated that no one misused or polluted water or deprived others of using it. There have also been references to the evolution of some rules and regulations relating to the protection of water and punishment for those deliberately polluting water. In addition, Prophet Mohammed had laid down some norms for the use of water. It was decreed by him that for vadhu and ghusul, only a bare minimum quantity of water be used. He himself set an example by using limited quantities of water for these purposes. Even in respect of irrigation, he is reported of have decreed that no more than an ankle deep water be used. As for the issue of equitable distribution of water, the prophet demarcated two types of zones—one which was called hima consisting of wells, springs or streams. These were meant to be reserved for collective use for consumption purpose, but bathing and washing were not permitted. The other category was called harims, reserved for growing trees and wood lands.

As Bible regards water as a gift from God emanating from his generosity, it insists that water be equally shared by all. In respect of the poor, Bible makes a special mention of water rights of the poor, indicating that the poor have a claim on water, a free gift of God which cannot be denied to anyone. The Bible specifically directs that water be offered to those who are thirsty and that at no time it be denied to anybody even if he or she is from the enemy camp. Since water was a scarce resource in the Middle East, it had the potential to create conflicts between different groups and communities. The Bible, under these circumstances, suggests that such people should resolve water disputes by entering into covenants.

In the Bible, there is a clear reference to water conservation practices which suggests that perhaps there was scarcity of water particularly for human and animal use. Subsequently, there evolved some belief systems which created awareness regarding protection and conservation of water. These belief systems are evident from the so-called proverbs and sayings. There were also references to water purification practices particularly in the Middle Eastern societies.

Awareness Through Belief Systems, Proverbs and Adages

Several beliefs helped conserve water and ecological balance in the past across the religious practices (Table 5). For instance, the suggestion that during rainy season, concentrating on fields and agriculture is important to make the best use of rains and natural irrigation is relevant. Belief systems like-prohibition of avoiding ceremonies that have a huge demand for water, during the month of 'ashada' as that was the period of the year meant for rejuvenation of aquatic bodies and breeding for aquatic life. Similarly, there were several belief systems, sometimes may be termed superstitious, that have helped protect water resources.

Table 5 Promoting awareness through proverbs

Islamic	Christian	Hindu
	• Pure water is the world's first and foremost medicine—Slovakian Proverb	• 'Don't live for even a single day wherever there won't be a banker, a Vedic laureate, a king, a doctor and a river' (Sanskrit)
	• A lake carries you into recesses of feeling otherwise impenetrable—William Wordsworth	• One tree for one household and one forest for one town (Kannada)
	• We never know the worth of water till the well is dry—Thomas Fuller, Gnomologia	• When someone has come to town should obviously come to fetch water (Kannada)
	• Water flows uphill towards money—Anonymous	• Why is the master's order required for fetching lake water? (Kannada)
	• The cure for anything is salt water, sweat, tears, or the sea—Isak Dinesen	

Water management and conservation are communicated through adages. Detailed references are made to proverbs, the distilled wisdom of old times and the aphorisms that stressed on the ethic of conservation, maintenance and augmentation of the water table, in their understanding and interpretation. Several proverbs and sayings exist in all Indian languages that explain the importance of water besides helping conserve natural resources in general some of which are listed. These proverbs and sayings are quite popular among the Hindus, as also the Christian community. However, we do not find any such references in the Islamic religion.

Technologies in Water Management

The basic question that needs to be started with is: were our ancestors aware of water cycle and hydrology? Happily, the answer to this question is: yes, our ancestors were aware of water cycle and hydrology as evident from the scriptures. The Vedic texts, which are more than 10,000 years old and many other scriptures such as Matsya purana, Bruhatsamhitaa and Arthashastra have valuable references to hydrological cycle which is very relevant to what is discussed in the current context. The application of technology in water management (architectural designs of storage tanks, hydrological cycles, meteorology, check dams, integration of different natural systems, etc.) across Hindu civilisations, dynasties, mythological accounts, Vedic literature, Puranas, Arthasastra, treatises, etc., is explained with a great degree of dexterity and clarity. All this highlights the scientific knowledge that prevailed in the past with regard to hydrology. Water purification practices were and are common across religions. Besides, they had specifications about the type of water, source of water indicating the quality of purity. Simple techniques were used and were popular among the people in purifying water where alum/herbs were used among the Islamic and Hindu societies. Other ways of purification like boiling and heating were also prominent. Besides, in Hindu scriptures, references are made to water getting purer when stored in certain containers. It is interesting and we can infer that simple practices to purify water could add to more consumption of quality water with least cost methods. Also one could cite the designs of water storage and conservation in support of the above claim. In Bible, there is a reference to the tunnel linking Jerusalem with the other side of the river. In India we can also cite the example of panchekki designed and constructed for grinding food grains in Aurangabad of the present day Maharashtra. The panchekki operated with water power.

With the available knowledge about water cycle and hydrology, our ancestors seem to have developed sustainable practices of storage and conservation of water. At the micro level, i.e. at the household level, there is a reference in Vedas to water being stored in copper containers in ancient India. The copper containers were preferred in the tropical countries like India for the simple reason that water stored in them would remain cool, whereas, in the Middle Eastern countries, water used to be stored in pots made of clay and animal skins for the same reason. Water in skin bags was preferred for another reason, i.e. it was convenient to carry water-filled skin bags on animals while undertaking long journeys in the deserts, particularly by traders and the army. As for India is concerned, water stored in copper containers is used for drinking purpose as also for washing idols and sprinkling water for cleansing the surroundings.

At the societal level, water storage and conservation received a greater attention in the ancient times. In Bible and Islam, there are many references to oases in the deserts with water stored for drinking and bathing purposes. In addition, there were natural springs also which provided water for consumption, and as such, it was stored in a proper condition by the community. Coming to India, the more frequently used devises of storage were open wells and water tanks constructed by the community. The latter were ubiquitous in South India, more particularly in Karnataka. These sources of water were largely used for irrigation purpose. And as for consumption purpose is concerned, there were traditional practices among some people, especially among Brahmins, who dug wells in their backyards and used water from these wells exclusively for drinking, cooking and bathing requirements by family members.

Since water is required throughout one's own life for varied purposes, people all over the world adopted some specific sustainable practices to conserve and to augment water resources. For augmenting water, the ancient people depended largely on rains. Rainwater harvesting was the usual practice for sustaining water resource under which water is stored in small pools such that it maintains groundwater table at a fairly higher level. Secondly, it was a practice among the people to plant and grow trees all around the residential and village areas as they were aware of the fact that the presence of tree cover attracted clouds and rains.

Ancient people followed traditional water sustainable practices, particularly during the biblical times; people in Palestine conserved and sustained water sources by digging and maintaining wells and springs so that they had access to water. For, according to them, water storage was a more effective way of sustaining water sources. Therefore, they preferred to dig and maintain wells and springs and store water in cisterns, public pools and aqueducts. An interesting piece of information is that for ensuring a continuous supply of water during times of siege by enemies, artificial pools were dug inside the walled cities which were often fed through a tunnel leading from a spring outside.

Largely, natural springs/rivers determined settlements. With respect to storage and distribution, immense engineering skills were showcased. Similarly, are instances of extensive skills in water resource engineering. They designed and constructed a variety of water structures and machines that are still fit for use. Prosperity of villages earlier came from water harvesting systems which gave an assured supply of water. Aqueducts were improved and modified into conduits. Underground canals were also constructed by Queen Zubaida in the eighth century AD for the towns of Makkah and Madina from 500 kms away with water stations in between. Such aqueducts are found also in India built in the sixteenth century by Moghul rulers through large networks. The aqueducts found are of different sizes. The scientific record of Al-jami bain al ilm wa-amal al-nafi f sina at al hiyal of Badi al zaman

Al jazri has a mention of 100 different mechanical devices. The author of the book himself was an inventor who gave siquia chain pumps with crank shafts, twin cylinder reciprocal suction pump driven by water wheel and automatic gates driven by hydropower-like water clocks and designed water supply system flush mechanism for hand washing and peacock fountain to mention a few. During the Christian era, canals, tunnels or aqueducts were designed to check evaporation of water almost completely, indicating advanced scientific management skills.

Awareness Regarding Water Cycle: Hydrology and Traditional Purification Practices

Across all the religious scriptures, civilisations and dynasties, we find enormous emphasis placed on technical designing of irrigation systems with storage and distribution as the focus (Table 6). The advanced irrigation systems designed in the past are functional even to this day. Well irrigation was prominent with several designs and types prevailing throughout history. Financing aspects were also addressed. Even common people had developed their own unique systems that were managed locally. It is interesting to note that in the first major human settlements, India started in the Indus Valley civilisation, along the bank of Sindhu River 5000 years ago. Excavations show the presence of private wells, common baths, pools, underground drainage systems, flood barriers and dock yards.

Water Governance Across Civilisations

Water governance focused on a lot of informal laws portraying moral responsibilities and also rights to access. This was prominent across all the three religions and provided guidelines to manage resources effectively (Table 7). We can also observe that equity and welfare concerns were the core of governance issues and were adhered to. However, there are references made to water conflicts in the Christian scriptures that were resolved amicably although at times battles were fought to establish control over rivers. There is also a reference to Palestine's people's conflicts over wells. Pricing of water is also seen as a taboo and to be made available to all.

It has been well documented that all over the world, civilisations flourished along the river banks. In the Middle East, Egyptian civilisation flourished along the Nile River, Mesopotamian and Sumerian civilisations along Tigris and Jordan rivers, and in India, civilisation flourished along the Indus and Gangetic valleys. As for India, the earliest civilisation, viz. Indus Valley civilisation, provides evidence to show how water use and management were efficient. As is expected, water was used by people for consumption and bathing purpose, but as an additionality, it was used for navigation purpose too. Thus, in Lothal—the ruins of Indus Valley Urban Settlement located in Gujarat—a harbour was built along the river for facilitating ships to carry

Table 6 Awareness on water technologies

Islamic	Christian	Hindu
Awareness regarding hydrology and water cycle		
• Allah, the almighty sends winds to cause rains on earth Quran (30.48, 7.57)	• Rain is believed to be a gift of god (Amos 9:6b, Job 36:27–28)	• Varahamihira speaks of water divining
• Percolation of water into layers of earth (Versus 23.18 and 19)	• Pouring from heaven, making earth fertile (Isaiah 55.10–11a)	• Hydrological cycle (Vedic texts)
	• All streams run in to the sea Ecclesiastes 1:7	• Kautilya refers to various kinds of clouds
		• Parjanya yaaga/cloud seeding technique developed in ancient India
Water purification practices		
• Treatment with alum/herbs	• Flowing water is recognised as living water, hence pure	• Treatment of water Ayurveda
• Sand, charcoal and pebble filtering of water at three stages		• Water was treated in specific ways to get specific qualities and used in therapies
• Boiling water		• Treating with alum/limestone/herbs
• Decantation		• Filtering with linen
		• Heating, exposing to sunrays/moon rays
		• Storing water in various metal containers
		• Techniques of zoo remediation and phytoremediation
Water distribution		
• Underground canals constructed by Queen Zubaida—eighth century AD—towns of Makkah and Madina with water stations	• Water covenants existed during biblical times	• Vedas recommend common source of water for all
• During Sultans (1191–1528 AD) and Mughals (1528–1857 AD) reign, aqueduct technology got introduced in India with many canals drawn for rivers in between	• In Mesopotamia (third century BC), canals were drawn for twin rivers	• Bhagiratha (Ramayana—first irrigation engineer—brought River Ganga—kingdom of Kosala
• Biggest network of canals was constructed by Sultan Firuz shah Tughlaq (1351–1386 AD)	• Assyria city got water supplied from deep wells	• Mahabharatha sage Narada advices Yudhishtira—irrigation works—adopted by the state

(continued)

Table 6 (continued)

Islamic	Christian	Hindu
• Wells—extensively constructed	• Mesopotamia had two farming systems—dry farming in the north (Assyria) and irrigated farming in the south (Babylonia and sumeria)	• Mauryas-Arthashastra, guidelines—construction of dams and references on water laws
	• Nirmud city wells were dug to 90 ft deep which provide water even today	• Guptas—irrigation works
	• Several steps for maintaining water bodies	• Cholas, Cheras and Pandya—several reservoirs and dams constructed—grand Cauvery anicut in Tamil Nadu
	• Roman Empire: water was brought to houses from channels through walled conduits and earthen pipes apart	• Pallavas and Gangas advanced irrigation systems including Chain tanks
	• Distribution/financing of aqueducts and Rome's water system are described in Rome's legal system	• Rajputs—development of best irrigation systems
		• Vijayanagara empire—advanced irrigation systems
		• Wodeyars of Mysore—great contributions to irrigation
		• Common people in village settlements developed local systems of irrigation and self-maintained them

Table 7 Water laws

Islamic	Christian	Hindu
Rights and moral responsibilities		
• Quenching thirst is a basic right of any individual	• Even an enemy should not be denied water	• Vedas recommend common sources of water for all
• Right of quenching thirst by men and animals is Chafa and irrigating lands is Chirb	• Only a villain will deny water to a thirsty person	• A blanket ban is put on the sale of food which includes water, education and medicine
	• Rich to be generous and help poor	• Food, medicine and education should be given to any deserving and needy free of cost
	• Awareness about ecology forms	• The sale of food, education and medicine will mark the degeneration of a society
Equity and welfare concerns		
• Quenching of thirst—allowed from private wells	• Strong emphasis—not to monopolise water	
• Extended to animals provided animals are not very large in numbers	• Poor/weak not to be deprived	
	• Having to pay for water is considered unjust	
	• Justice forms a main feature of Christian principles	
• 1870 Ottoman Empire codified laws of water named Mejelle. Mejelle calls water an unsalable commodity	• Rome's legal system describes distribution and financing of aqueducts and other pieces of Rome's water system	
• Ownership, pricing and sale of water came into existence with the development of civilisation	Bund construction—flooding regulation	

goods and sail into the Arabian Sea en route to Middle East. This harbour indeed exists even today though it is not being put to use. The interesting point is that the ships would be anchored at the harbour which is like a pond, and as goods were loaded into the ships, they would sink a bit on account of the heavy weight of goods. As the partially sunk ships could not be manually taken out of the harbour, the people had devised a method of floating the ships out of the harbour. And that was by constructing a dam on the upper reach of the river where water was stored and whenever the loaded ships were ready to be floated out of the harbour, the crest gates of the dam would be opened up to allow the water to gush out such that the ships were pushed and carried away from the harbour into the river flowing into the sea. In other words, people of the time had understood the power of water and knew how to put it to their use.

Water management practices during various dynastic rules in the past make for an interesting reading. If we go back to the Christian era, in the Middle East, the main occupations were animal husbandry, especially rearing of camels and sheep, and agriculture and business. For the first two occupations, water was considered essential. Hence, water storage became essential for use by animals and crop cultivation. It was realised that storage of water was one thing and channelising of water for use in agriculture was another. Therefore, water management became an important dimension of water use in the sense that stored water was to be efficiently distributed on a day-to-day basis. For this purpose, the Kings and Queens dug up canals for facilitating the flow of water to lands owned by farmers. Besides, there was this case of Queen of Haroon-Al-Rasheed of Baghdad who got a canal dug up for bringing water to Mekkha where water scarcity was experienced by pilgrims. In all such cases, gravitation was used as a means of conveying water from the upper to lower region. In India also, right from the days of Indus Valley civilisation to the present, one can observe a network of canal systems put in place to meet the water requirements of agriculturists as also of people in general. Here also, since the days of the occupation of the Indian people right from the Dravidians of the Indus Valley and the Aryans of the Gangetic Valley down to south covering Pallavas, Cholas, Gangas and Kadambas and Vijayanagara Kings, the rulers used the gravitational force to convey water from one point to the other for agricultural purpose. Those of the settlements living along the river banks lifted water using the traditional methods such as yetha in South India, Persian wheel in Middle East and so on.

The water stored as well as channelised was to be distributed equitably among all the members of the settlements. In order to ensure an equitable water distribution, the rulers employed local workers especially from the lower caste who went by the name of neeruganti in Karnataka, for example. The working of this system required special regulatory mechanisms and rules to be framed by the local governments. During the early years, these rules were informal in nature, but more recently, they have been formalised by the local governments. The basic purpose of framing formal rules is to reach water even to the fag-end lands.

There have been instances of water use conflict too. There is a reference in the Christian scriptures to such conflicts erupting, especially between people of different settlements which, of course, today have been translated into interstate and international water disputes. During the early period, such conflicts related to water use were resolved through consensus and a policy of give and take. Today, however, such resolutions have become more formal in terms of taking the disputes to courts of law for legal remedies, causing not only undue delays in getting justice but also heartburns on the part of litigants.

All the scriptures across religions have emphasised on effective ways to improve water resources management. Exceptions may be seen in celebrating water festivals or variations in beliefs systems in perceiving water as divine element to be worshipped or vice versa, but the emphasis on sustainable water resources management with focus on augmentation, protection and conservation is evident commonly across scriptures. All the scriptures respect water and part of the numerous rituals and practices in people's lives and hence act as catalysts and influence informal laws. Hence, good water governance in the ancient past had foundations based on the distinct local culture and unique history, state and its institutions.

References

Bertus H, Reijntjses C (2007) Moving world views—reshaping sciences, policies and practices for endogenous sustainable development, ETC/COMPAS, P.O. Box 64, 3830 AB, Leusden, Netherlands

Manthrithilake (2012) Presentations during the 3rd Run of the Temasek Foundation Water Leadership Programme (TFWLP), scheduled to commence from 12th to 23rd November 2012 at the Lee Kuan Yew School of Public Policy (LKY School). National University of Sigapore's Bukit Timah Campus in Singapore

Oestigaard T (2009) Chapter 1: Water, culture and identity: comparing past and present traditions in the Nile basin region. BRIC, Bergen

Stewart M (2008) C E 397—Transboundary water resources, 10 Apr 2008

Websites

http://exnora.in/pg/about/read/about

http://pib.nic.in/newsite/erelease.aspx?relid=83056

http://portal.unesco.org/culture/en/ev.phpURL_ID=2450&URL_DO=DO_TOPIC&URL_SECTION=201.ht

http://timesofindia.indiatimes.com/city/bangalore/Karnataka-State-Pollution-Control-Board--urged-government-on-eco-Ganesha/articleshow/21688583.cms

http://timesofindia.indiatimes.com/city/lucknow/Polluting-Gomti-regularlyreligiously/articleshow/29786770.cms

http://www.constructionweekonline.com/article-19723-water-saving-mosque-initiative-wins-5000-award/

http://www.satyamevjayate.in/Dont-Waste-Your-Garbage/EPISODE-3watchvideo.aspx?uid=E3-EV-V1&lang=hindi

http://www.speakingtree.in/spiritual-slideshow/seekers/faith-and-rituals/worlds-1st-flower-reprocessing-machine-sai-templedelhi-289892

http://www.unep.org/NEWSCENTRE/default.aspx?DocumentId=617&ArticleId=6491

http://www.waterculture.org/Water_Ethics_Charter.html

http://www.waterculture.org/Water_Ethics_Network.html

Policy Perspectives: Water and Scriptures

A. Ravindra, K.V. Raju, and S. Manasi

Introduction

Water is one of the five basic elements of nature (the others being earth, air, fire and ether) which constitute the primordial substance from which the universe came into being. No wonder, people began to settle down near rivers, a major source of freshwater, and civilisations were born—Mesopotamian, Egyptian and Indus Valley—and Chinese civilisations were gifts of the Tigris and Euphrates, the Nile, the Indus and the Yellow rivers, respectively. What must have been abundant to begin with, when the human population was small, this natural resource gradually became something in short supply. With a continuous growth in population and human activities, the need arose to store water for purposes like drinking, washing and irrigation. In times of shortages and scarcity, people did not hesitate to quarrel for water. Thus, what was a common resource once became a problematic resource later.

Water conflicts were not unknown even in ancient times. The story goes that over 2500 years ago, the Sakhyas and Koliyas, two tribes in northern India, waged a bitter war over sharing the waters of the river Rohini resulting in bloodshed. When the Buddha came to know of it, he intervened and brought the two sides together, putting an end to the long-drawn battle.

In modern times, the water requirement for various uses like drinking, sanitation, agriculture and industry has increased manifold. During the twentieth century, the

A. Ravindra (✉)
Chairman Board of Governors, Institute for Social and Economic Change, Bengaluru, India
e-mail: dra.ravindra@gmail.com

K.V. Raju
Policy and Impact, Asia Regional Program, International Crops Research Institute
for the Semi-Arid Tropics, Hyderabad, India

S. Manasi
Centre for Research in Urban Affairs, Institute for Social and Economic Change (ISEC),
Bengaluru, India

K.V. Raju, S. Manasi (eds.), *Water and Scriptures*,
DOI 10.1007/978-3-319-50562-6_7

global population increased fourfold to more than six billion, while the water withdrawn from natural freshwater ecosystems increased eightfold during the same period (Gleick 1998). The growing demand-supply gap has created serious concerns regarding the availability of water in adequate quantity and quality for different sections of the population, particularly in developing countries.

In this chapter, we examine policy issues related to water management in India with reference to ecological and cultural dimensions.

Key Issues in Water Management

India has been classified as a water-stressed state, i.e. a condition where the per capita availability of water is less than 1700 cum. The total utilisable water in India is estimated at 1123 BCM (Ministry of Water Resources), while another estimate puts it at 654 BCM (Narasimhan 2008). The projected water demand comes to 1093 BCM by 2025 and 1447 BCM by 2050 (MoWR 2006). The per capita average annual availability of water is expected to decline from 1816 cum/year in 2001 to 1340 cum/year by 2025 and 1140 cum/year by 2050.The UNICEF, in its report 'Water in India: Situation and Prospects', (Prakash Anjal et al. 2013) has identified the following crucial issues faced by the water sector in India:

- Erratic distribution of rainfall, often leading to floods and droughts
- Water use inefficiency
- Unregulated groundwater extraction
- Water pollution
- Decreasing water quality due to poor water management laws, interstate river disputes, growing financial crunch for development of resources and safe drinking water
- Inadequate institutional reforms and ineffective implementation

Water in the Context of Ecology and Culture

Several problems relating to the water sector may be attributed to the lack of awareness about historical, ecological and cultural processes of a country or region. Increased run-offs, soil erosion, flooding, aquifer depletion, water quality degradation and climate change are some of the problems associated with such neglect. The ecological consequences of neglecting the water needs of freshwater species and ecosystems have been tragic (Pringle et al. 2000; IUCN 2000).

There is also a cultural dimension to water, people's beliefs, values and perceptions playing an important role in influencing the use of water. 'The centrality of water in human life has made it arguable that the ways in which it is conceived and valued, understood and managed, used or abused, worshipped and desecrated, are influenced by the cultures' (World Water Council 2003). Martin (1991) pointed that people share certain beliefs and values, compete for some objectives and differ in ideas about power and how it should be exercised. Rules to be complied by society are set and central organ of the ruling power in that society also from explicit/implicit cultural agreements about what is acceptable and desirable. The state cannot be separated from culture, and governance cannot be considered without reference to its cultural context. Inclusion of the principles of ecological sustainability and cultural knowledge in water policies would undoubtedly enable a more efficient management of water resources. Let us look at the constitutional and current water policy framework in India.

Federal Structure

Under the Indian Constitution, water is a state subject, but the parliament has an overriding power over regulation of interstate rivers and river valleys in the public interest. Exercising this authority, the parliament passed the Inter-State Water Disputes Act, 1956. The central government's role is thus confined to providing a legal framework for managing interstate rivers and resolution of interstate water conflicts. It can also frame broad policy guidelines on matters relating to water management keeping in view the geography of a vast country like India and its physical and cultural diversity. Art 262(2) of the constitution bars the jurisdiction of the courts to adjudicate on river water disputes. In practice, however, states have, on many occasions, approached the Supreme Court against the orders of the tribunals constituted for dealing with interstate water disputes (Guhan 2007).

The states may formulate their own policies and along with local governments are responsible for the implementation of plans and projects for drinking water and irrigation. Water supply to rural and urban areas is the responsibility of the Panchayats and the urban local bodies as per 11th and 12th schedules of the constitution. The ecological, economic, social and political factors at play have made water governance a complex process.

National Water Policy (2012)

The National Water Policy recognises water as a scarce natural resource 'fundamental to life, livelihood, food security and sustainable development' and as a 'sustainer of life and ecology'. It further notes that low consciousness about the scarcity of water and its life-sustaining value has resulted in its 'mismanagement, wastage and inefficient use, as also pollution and reduction of flows below minimum ecological needs'. The policy accords importance to the role of the community in water management and states that the community must be sensitised to adapt first to the utilisation of water as per local availability before providing through a long-distance transfer. Taking note of the likely impact of climate change on increasing the variability of water resources affecting human health and livelihoods, it has emphasised the need to enhance the capabilities of the community for adopting climate-resilient technologies.

National Environment Policy (2006)

According to the National Environment Policy, India's freshwater resources 'comprise the single most important class of natural endowments enabling its economy and its human settlement patterns'. It further advocates the promotion of integrated approaches to the management of river basins in order to ensure the maintenance of adequate flows and, in particular, in-stream ecological values and adherence to water quality standards.

State Water Policies

Various states in India have come up with their own water policies. The following issues may be considered crucial in the context of policymaking and governance: ecological sustainability, equity, economic efficiency and holistic water management.

Ecological Sustainability

In the first place, the water-related ecological objectives must be quantitatively defined. This would involve estimating the ecological flow requirements in each river basin and linking them to water management. Hydropower engineers argue, mainly from the energy planning perspective, that 10% of an ecological flow would be enough, while the Wildlife Institute of India has suggested 20–30% for different seasons. The Centre for Science and Environment has recommended 30% e-flow for 6 months (May to October) and 50% during the remaining 6 months (November to April). The Inter-Ministerial Group headed by BK Chaturvedi, former Planning Commission member, has proposed 25% for 8 months and 30% for 4 months.

The critical issue, from a policy perspective, is planning for hydropower generation. The main argument in favour of hydropower projects is that apart from being a clean source of power, the cost of generation is low, and power tariffs can be kept low. However, this is a lopsided view. Maximising power generation from a river cannot be the sole objective. Sustainability should be considered an important part of the planning process. As such, the appropriate e-flow requirements of a river should be taken into account while estimating the potential of hydropower generation.

The increasing extraction and use of groundwater over the years has emerged as a matter of serious concern. The share of groundwater in the net irrigated area in India has increased to 58% as against 28% of surface water. The number of wells across the country is estimated at about 19 million (tube wells and dug wells). Groundwater levels are sharply declining on account of reckless exploitation. The average bore well depths in states like Karnataka and Maharashtra have reached more than 1000 ft, touching 1400 ft in some regions.

Groundwater is one major resource which is mostly unregulated in both rural and urban areas. In rural areas, the policy of providing cheap power to the farmers for pumping is not only leading to reckless exploitation of groundwater but also wastage of water and power. In urban areas, the problem is related to a shortage of drinking water on account of rapid urbanisation. Attempts to regulate the use of groundwater have not met with success. For instance, Karnataka enacted the Groundwater Regulation Act in 2012, but its implementation has hardly been effective.

The increasing pressure on water resources, particularly for urban and industrial purposes, has caused a severe level of water pollution affecting the health of water bodies and human beings. Growing urbanisation not only makes heavy demands on higher quantities of water supply but also leads to generation of huge quantities of waste water and sewage (Corcoran 2010). India's towns and cities are estimated to generate 38.2 bn litres of sewage every day, while the installed capacity for treatment is only 31%. According to the Central Pollution Control Board, class 1 and 2 cities along the Ganga river alone generate 2.7 bn litres of sewage every day, but the treatment capacity exists only for 1.2 bn litres, thus putting back 55% of the untreated sewage back into the river. The level of coliform bacteria (used as a measure of river health) ranges from 25,250 at Rudraprayag in the Himalayas to 58,000 in the holy Varanasi to a high of 6,53,083 in Dakshineshwar near Kolkota, compared to the ideal level of 5000 per 100 mL. Similar is the story of the state of most of our rivers and other water bodies across the country.

Equitable Distribution of Water

The competing uses of water have led to inequities in distribution—between agriculture and industry, urban and rural areas and the rich and the poor. A key objective of water policy is to ensure an equitable distribution across sectors, regions and different sections of the society. Allocation of water for different uses therefore needs to be done judiciously. Apart from drinking water, agriculture is accorded a high priority in water policy. However, increasing industrialisation and urbanisation are making heavy demands on water. In fact, a highly industrialised state like Maharashtra had given a high priority to industry over agriculture in the allocation of water in 2003, but the government was forced to reverse it in 2011.

Inequity can be seen even within a sector. In agriculture, for instance, crops like paddy and sugarcane consume huge quantities of water compared to dry crops like legumes and oilseeds. While the right quantity of water must be made available to different types of crops, the problem arises when water-intensive crops are grown in areas earmarked for dry crops in irrigation zones. According to D.M. More, former director general of Maharashtra Water Resources department, sugarcane alone consumes water equivalent to the total storage capacity of all dams in Maharashtra. This not only amounts to a misuse of water but is also detrimental to the arid soil. Thus, water available to agriculture is not used prudently.

In urban areas, particularly in large cities, spatial diversity causes an inequitable distribution. Assessing a city's water requirements purely on the basis of per capita supply norms, i.e. litre per capita per day (lpcd), as is done by water utilities, is rather simplistic. Studies have shown that the lpcd data hide the facts relating to water losses caused by leakages and spatial variations. In Bengaluru, for instance, the per capita supply, according to official estimates put out by Bengaluru Water Supply and Sewerage Board, is 120 lpcd. However, this does not take into account the water leakage amounting to 45%. Moreover, some of the peripheral areas of the city do not have access to piped water supply and depend on groundwater. The spatial variations range from less than 50 lpcd to over 200 lpcd. As the city sprawl expands, the problem is not confined to supply of water alone but to sanitation and waste disposal as well, especially in underdeveloped areas and slums. A recent study has argued that the traditional 'planning' exercise in urban water utilities needs a serious overhaul and has proposed an urban metabolism framework for water. 'When both water flowing into the city and the waste emptying out of the city are properly accounted for, a social metabolism framework allows for combining questions of justice and ecological sustainability' (Mehta et al. 2014).

Economic Efficiency

Huge investments are made every year in irrigation projects, drinking water supply schemes and sewage treatment plants. The returns, however, are hugely disproportionate to the investments made both on account of shortsighted policies and inefficient use of water. Needless to say, this has serious economic implications. The policy of providing cheap power for agricultural use has resulted in an excessive use of water without the users paying for it. Most of the states have adopted this policy for the officially stated reason of subsidising poor farmers but, in reality, to cater to the vote bank and extending the benefit to the rich farmers as well. In Punjab where groundwater meets three quarters of the farming needs, free power is considered the main reason for the state's revenue deficit and fiscal crunch.

Investments in water assets—watershed development, soil conservation schemes, etc., particularly through employment guarantee scheme—are ill planned reflecting a great deal of inefficiency. Trees are planted but are hardly well protected. Under MNREGS, the focus is more on providing employment than creation of productive assets. Out of six million waterworks taken up under this scheme, 60% are said to be incomplete with only 8% of the people availing of full benefits (Mahapatra 2008).

Economic efficiency demands a full value for the use of water through physical as well as managerial measures. But water is also a public good, and it may not be appropriate to expect full recovery of costs in financial terms, especially in respect of highly capital-intensive works. However, water pricing must at least ensure a recovery of the cost of maintenance besides including an element that does not incentivise wastage of water.

Integrated Water Resources Management

According to the National Water Policy, water must be managed as a common pool community resource, held by the state under the public trust doctrine. This is possible only if, in the first place, the government itself adopts an integrated approach to managing water resources. What we find in reality is compartmentalisation of the water sector, both at the central and state levels. At the centre, there are separate ministries for dealing with irrigation (Ministry of Water Resources) and drinking water supply to rural areas (Ministry of Rural Water Supply) and urban areas (Ministry of Urban Development). Similar is the structure in the states, except that the water resource department is further split into major and minor irrigation. Lack of a unified perspective in planning, management and use of water resources coupled with institutional inadequacies is acting as a serious barrier to effective implementation at the ground level. Droughts and floods which are common features in India must be viewed as part of water management than just a one-time disaster management. A phenomenon witnessed in the recent years is the extreme and variable rain events. The heavy and unseasonal rainfall that devastated Uttarakhand in 2013 occurred in the month of June, which was not the beginning of monsoon in that region. The resultant floods and landslides trapped about lakh pilgrims, leaving over 5000 dead. Scientists may link such events to climate change, but the role of man-made infrastructure interventions in the hilly and forested regions cannot be ignored. In urban areas, heavy rains lead to flooding of streets and clogging of storm water drains. In coastal cities such as Mumbai and Chennai, normal life gets paralysed during certain days of the monsoon. The impact of climate change on water resources with reference to variability must be studied across different regions, and state governments must develop strategies to combat the extreme weather events.

Cultural Viewpoints

An understanding of the cultural dimension of water use would help sensitise communities to adopting efficient water management techniques. There is something to be learnt from the traditional wisdom of the Indian people in conserving water. Rainwater harvesting which essentially means collecting water when and where it falls is an age-old practice, and different ways were found to store the water collected. The thousands of tanks created in Karnataka are testimony to the wisdom of the rulers and people of the state in water conservation techniques. Johads or soak pits dug to conserve and recharge groundwater is a tribute to the native intelligence of the people of Rajasthan. The chapters so far have extensively discussed and throw light on the beliefs, traditions and practices of people with regard to water in the context of religion, region and culture. Concepts such as water conservation, justice, equity, purity (quality), etc. have been prevalent over the centuries. What is important is to communicate them to the people in accordance with the requirements of modern times. Today we have the advantage of technology which has enabled mass production and transfer of water over long distances. But this should not preclude efforts to first utilise water as per local availability before resorting to long-distance transfers. Local cultural models can be evolved drawing on past experiences combined with modern ideas which can appeal to the people. Modern technology can also be used to fine-tune the cultural rhythms of communication. Even to this day, cultural traditions in the form of rituals prevail prominently; however, making the rituals a part of augmenting, conserving and protecting water for sustainable ways has to be embedded into our systems through policy. This could save costs as well as water.

Challenges in Convergence Culture

The big challenge is to achieve convergence of policy, technology, social sciences and governance. This would essentially involve the integration of 'ideational domain' (beliefs, values and perceptions concerning water), 'organisational domain' (how water resources are harnessed) and 'institutional domain' (structures, rules and procedures) for an effective management of water resources. Cultures are dynamic and change in response to the external influences on them. In pursuing what one society has agreed to be good governance, there will be some agreement as to what is considered 'good'. Although, it seems simplistic, 'ensemble of ideas' included in cultures constitutes a code that establish ethical orientation of a particular society and relates to its institutions and its structure of power. This ensemble is not static or closed. Hence, the phenomenon of 'forgetting the past' may, perhaps, be on account of a variety of domineering influences. These include significant and overwhelming alien influences that succeeded in literally wiping out a substantial aspect of tradition and ecological culture. Distortions, for instance, in the context of India, and language of the texts of the past mostly being Sanskrit have led to an improper interpretation of scientific principles and the assumption that the language is Brahminical, and hence it was not widely taught, used and understood by all sections of the society coupled with the attitude of the learned class in the past who also kept the knowledge encrypted and confidential. Similarly, deviations, misconceptions and misinterpretations for political gains and power led to divide in society and disintegration, like, for example, focus towards centralised water supply, scientific literature taken away, growing use of subsidised energy to exploit deep aquifers and reliance on big dams coupled with overall neglect of prevailing technologies based on ancient wisdom.

Perpetuation of dominant dogmatic practices, against which effective safeguard measures did not exist or last for long further increasing with apathy by policymakers towards water harvesting structures post-independence and the lack of innovative ways to deal with water related issues, has led to the present crisis. Subsidised supply of fertilisers has reduced the dependence of farmers on tank silt. With many of the tanks being encroached upon for farming, sand mining, expansion of city, waste dumping including industrial effluents, etc., tanks and rivers have virtually become dump yards. Further, a declining interest on the part of communities in nurturing various traditional water harvesting systems further affected the traditional system. The loss of traditional systems has resulted in acute and chronic consequences. Currently, with emphasis on centralisation and modernisation of water management systems; water is not supplied regularly by the governing bodies and, even if supplied, is not at all in proportion to the needs. The groundwater levels have depleted, and the quality of the water in general is affected extensively.

UN reports suggest that there are around 300 potential conflicts over water around the world, arising from squabbles over river borders and the drawing of water from shared lakes and aquifers. Internationally, water conflicts may stem from the drive to possess or control another nation's water resources, thus making water systems and resources a political or military goal. Internally, within a nation, an inequitable distribution and the use of water resources, sometimes arising from a water development project, may lead to development-related disputes. In India, governments are faced with frequent protests over the allocation of river waters between states and protests by farmers over the allocation of waters to industries. In the latter case, governments face the dilemma of having to build large dams to meet development needs on the one hand and protecting the interests of farmers, on the other. The environmental concern also comes into play. We have the task of making difficult ethical and political decisions, to reconcile the often conflicting goals of technological progress, environmental preservation and human values such as justice and freedom.

Cultural Trends and Implications

Two megatrends have emerged (1) in urbanisation a cause of 'capitalist modernity' and producing a broad process of convergence in the cultures of the world. (2) Centre of this development the culture capitalism promoting consumption or consumerism (Tomlinson 1991). Hall (1997) describes this homogenisation as a characteristic of what he terms global mass culture. Producer and international connected forces of production use the seductive key of advertising and instruments in forms of clothing and new technology, to hook the consumer (Appadurai and Breckenbridge 1990). All these affect the ways in which people live and affect usage/abusage of natural resources. Good governance, through the representatives, should present alternatives to this economic dominated mode—'what various societies want and can have'. Characteristic of modernity—process across various spheres of life—becomes progressively separated, for example, separation of domestic work or reproductive roles from productive role, church from state, law from morality and state from society. Good governance seeks to bring back the possibility of reconciling these separations through social policies. State, in governance terms, should be closer to society through better representation.

Learning's for the Future

Given the changing contexts, drawing lessons from the scriptures for the current and future water resource management is seen as one important dimension and was one of the main aims of studying them across religions. However, we ought to acknowledge that there has been a paradigm shift during the last five or six decades. Having reviewed the papers on water, based on the messages from the scriptures, we identify lessons for the future, which for sure needs the modern twist given the newer challenges, however, based on the approaches/principles of the past. The lessons for future should cover all the dimensions of water use such as augmentation, storage, protection, conservation and distribution. We know that freshwater constitutes a small portion out of the total available water on the earth and challenge due to increasing demand. Hence, the way we look at water and how we use it have to undergo a major change.

As all the scripture (Islamic/Biblical and Hindu) perspectives suggest, our individual attitudes and practices should reflect a conscious recognition of water as god's gift to humanity, almost as a sacred element. Sacred elements, though used for human benefit, are revered and preserved, not wasted. Water is precious because it is life giving. With increasing demand, people should, therefore, be careful and considerate in the usage of water in our daily lives. These aspects have to be captured and inculcated into our governance systems for proper water management.

Storing Water

The need for storage of water has become more urgent because one cannot today completely depend upon rains for all our water requirements. This is because rains are confined to a specific season and that too to a small duration. The question of 'what should we do' during the rest of the seasons will stare at our face. Perhaps, this question was also present during the ancient times because the religious scriptures refer to storage practices during those days. Normally, water was stored by constructing small barrages, digging wells and building traditional water tanks. Of course today, considering the quantum of water required for consumption and irrigation purpose, the storages—whatever be their kind—have to have large capacities. That is why we go in for major irrigation projects like dams across rivers. In the rural areas, even today, we take up construction of new irrigation tanks and de-silting of old ones. Of course, on account of the fact that groundwater table is depleting; almost everywhere open wells have nearly become defunct. However, lessons can be taken from the practice followed in the ancient times for improving groundwater levels by recharging groundwater through harnessing rainwater. This is being done by the state today by undertaking watershed development programmes, while at the same time, the state government has been insisting upon rainwater harvesting by households in some urban settlements. But since these measures are implemented on a selective basis, the outcomes of which do not appear to be very satisfactory.

Protecting Water

In the past also, perhaps there was a need for protecting water against water pollution as evident from the directive found in the religious scriptures that water be not polluted and that the polluter be punished for such an act. Since a moral force was exerted upon people by the religious scriptures, people respected the word of religion and accordingly abstained from dirtying water. But today since moral pressure does not work, there is a need to use legal power for preventing people and industrial enterprises from polluting water. Awareness about the invaluable nature of water and other resources has to be spread widely so that everyone regulates oneself and involves in protection and upkeep of resources voluntarily. This could be achieved through a proper religious education through scriptures.

Conserving Water

Regarding conservation of water, there are some references in the religious scriptures to water conservation methods. For instance, the Islamic scriptures state how Prophet Mohammed set an example to conserve water by not wasting water. Stringent regulations and policing may result in a fruitful impact, but a great sense of responsibility takes birth among the rule-maker and the ruled when the rules and regulations are internalised thoroughly, which has been the philosophy that resulted in effective enforcement in the past. Yajurveda says 'dedicate your lifespan to perform the best tasks, dedicate your breathe to perform the best tasks, dedicate your eliminatory power to perform the best tasks, your assimilative power to perform the best tasks, your blood to perform the best tasks, your sight to perform the best tasks, your audibility to perform the best tasks, your mind to perform the best tasks, your voice to perform the best tasks, your soul to perform the best tasks, the best of your powers and actions to perform the best tasks'[1]. Although, the Vedic scriptures have put a blanket ban on sale of food, education and medicine, it may not be feasible in the current contexts. However, it does highlight the moral approach they adopted in promoting natural resource management. With respect to water, a form of food or

[1] आयुर्यज्ञेनकल्पतां प्राणो यज्ञेन कल्पतामपानो यज्ञेन कल्पताम् व्यानो यज्ञेन कल्पतां चक्षुर्यज्ञेन कल्पतां श्रोत्रं यज्ञेन कल्पतां मनो यज्ञेन कल्पतामात्मा यज्ञेन कल्पतां यज्ञो यज्ञेन कल्पताम्.... यजुर्वेद:/aayuryajnenakalpathaampraanoyajnenakalpathaamapaanoyajnenakalpathaamvyaanoyajnenakalpathaamchakshuryajnenakalpathaamshrootrmyajnenakalpathaammanoyajnenakalpathaamvaagvajnenakalpathaamaatmayajnenakalpathaamyajnoyajnenakalapathaam..... Yajurveda.

food itself as described by Vedas, water should never become a priced good in the hands of private pliers in the least. Such an act would land the poor in miseries of drought amidst abundant availability of water leading to potential water wars.

Sustainable agricultural practices are discussed in scriptures and also across civilisations that make efficient use of non-renewable resources and on-farm resources, where appropriate, natural biological cycles and controls are employed minimising wastage, reducing consumption especially of water. Employing sprinkler and drip irrigation techniques and maximum dependence on natural irrigation (rainfed farming), raising of local crops using the zuari or natural or locally available seeds, adhering to the specific geographic conditions and reusing and recycling of water and all other things that could be recycled can help save most of the 70% of the pure water spent on irrigation. Usage of organic and natural fertilisers has also proven to reduce the amount of water required to raise crops. In the rural situation, water can be saved using the irrigation scheduling method. It is economically beneficial to producers as it helps better manage the available water and increase water use efficiency, while reducing pumping costs. Applying only the amount of water needed for a particular crop can reduce input costs without sacrificing production. In addition, over application of water can potentially affect crop root growth and the resultant yield levels. In the water context, the papers have highlighted immense knowledge and relevance of the ancient scriptures to our current context, particularly the wisdom that the ancient civilisations followed. Conservation and respect to nature was a part of their culture and daily life. Hence, this is one of the important lessons that need to be incorporated into the governance framework.

Equitable Distribution

As the scriptures clearly indicate, water is God's precious gift to all. Therefore, it is important that we recognise the legitimacy of the common claim over water by all. This calls for water policies that ensure an equitable distribution of water resources such as river waters among states and regions in the country and an easy access of clean drinking water to all sections of the society. The approach of our governments to river water management should be realistic and oriented to population needs. Our public leaders should function more like statesmen with long-term goals and general good in mind instead of appeasing parochial sentiments and agitations. According to the Bible, governments are called into being to provide security, to maintain order and to promote justice and development. The governments, therefore, have a greater obligation to provide clean drinking water to their citizens. In order to defray the heavy costs involved in fulfilling this responsibility, governments are resorting to hand over the basic control of water systems to multinational companies. Today, water is the third largest industry after electricity and oil. The argument made in favour of privatisation is that making clean water available would reduce large-scale deaths due to water-borne diseases. But the moral and ethical questions raised against privatisation deserve an equal consideration. If potable water comes only at a cost, should those unable to pay for it do without it? Considering the large number of poor people in the world, by pricing water, are we not risking the onset of water wars as is already the case with oil? If potable water can come only with a cost, should it not be the responsibility of the governments to subsidise the costs for the very poor? A better alternative to experiment with is to promote low-cost technologies and treatment plants owned and managed communally by neighbourhood communities and villages. This will ensure an easy access to clean drinking water as well as inculcate values of self-reliance, equity, social justice and eco-justice.

Although water crisis was not severe during the biblical times as it is today, biblical history was made around Palestine that had chronic shortages of water supply. Yet, biblical people held the view that water was to be shared with all and that conflicts over water be resolved with that perspective in mind. Today, water shortage in the urban areas is often being managed with supply augmentation options such as own wells and in-house storage, water purchase from private tankers, inter-household water sharing and privately organised small-scale water transfer from irrigation. Water is brought from distant and multiple-use water sources which gives rise to conflicts over allocation for industrial and agricultural purposes. Some of these options are helpful, but an unrestricted digging of wells in urban areas can prove counter-productive. Groundwater is a vital resource, and everyone should benefit from a sustainable supply of clean groundwater. Efforts should be doubled to retain groundwater levels. The local governing bodies should strictly implement laws restricting digging of too many bore wells.

Water Pricing

A few argue for a more rationalised water pricing policy and exhaustion of exploring local level supply augmentation options. According to them, although local level supply augmentation options cannot, by themselves, solve urban water deficit altogether, their exhaustion is a necessary condition for market-based inter-sectoral water transfers to be free of the damage to the incentive environment facing urban water sector. They call for reforms in the pricing policy through legal and institutional changes to fully exhaust local level options and to economically justify macro-level options. In cases of conflicts, governments should ensure both the economic feasibility and geological feasibility. At the same time, the problem of large-scale displacement of people should be minimised. If, however, displacement is unavoidable, plans should be made well in advance to provide adequate rehabilitation.

Based on an analysis of the directives of the religious scriptures and the practices of water conservation by our ancestors, we can learn some lessons for the future. These lessons may be delineated for (1) individuals, (2) communities and (3) state or government. The chapters have brought to fore several suggestions drawing from scriptures, indicating the role of individuals, community and state, which needs to be packaged to suit local contexts. In this context, broad lessons that were also emphasised was (1) behavioural change across individuals through education and awareness, (2) promotion and adoption of low-cost and decentralised technologies, (3) more active community participation and (4) state to promote reuse, ensure equitable distribution, promote conservation and provide incentives. Tables 1, 2 and 3 that follows is an attempt in that direction.

Some Observations

Ways of life and religion helped a lot in ensuring that certain things, especially natural resources, were protected besides paving the way for laws to ensure the same. While it can be said that religion and religious practices have continuously played an imperative role in water conservation, the lessons that we have learnt need to be put into practice more efficiently. Steps that were put into practice in olden times need to be revived or brought back in a way that is conducive to the present scenario. Religion, as we are all aware of, plays an important part in moulding man's behaviour and attitude towards nature. The catch is how to put to use the religious principles and practices in such a way as to ensure that the very spirit behind it is imbibed and put to good practice. Only if we are able to imbibe such a spirit can the true value of the same be put into practice. The basic objective should be to ensure that the principles and practices that were prevalent and which had a religious backing to them translate into something concrete as law is and thereby exert some sort of a commanding power. This power that it exerts is to ensure that people follow it and, if not, is subject to penal sanctions.

Religion can be found to have an impact on man that law is not capable of exerting. Law has its limitations, owing to man treating it as an external factor, a factor that is imposed on man by reason of him being a subject of a particular land. It is this aspect of religion that helps it big time to score in an area where law has not been able to make its mark. Stamping its authority, religion, it can be said, has been highly successful in checking the actions of man and creating in him a sense of togetherness with all things present in nature in general and all life in particular. It is this aspect that religion possessed that made it a worthy contender to instil the duty to protect environment in man. Religion, it is seen, had, as a result of umpteen indirect acts, been able to protect the environment and water in particular, as can be seen from the preceding works. When an innate sense of duty is being created in man, mandated by the tenets of religion, he is bound to observe it fearing divine sanctions. This sanction, as pointed out earlier, has a higher level of acceptance amongst people than the sanction of law. To a believer, punishment from above is in all probability more severe than punishment which has been provided for by the state-made law. When religion preaches such a concept of man being part of nature and imposes on him a duty to take care of it owing to his superior status, it is imperative that it has a strong effect in terms of shaping man's actions. This was something that was part and parcel of religion and cannot be achieved by anything else, let alone law, an external agency.

Table 1 Learnings for individuals

Individual—focus on behavioural change—responsible use and adopt technologies	
Augmentation	• Mandatory rainwater harvesting
	• Citizens to be made responsible and accountable
	• Tree planting at houses/road side and maintenance by household members—a moral binding
	• Awareness creation to be made part of culture
Storage	• Rooftop water harvesting
	• Use of herbs—tulasi, lavancha and alum to improve water quality
	• Replace plastic containers with bronze, copper and earthen containers to maintain purity
Protection	• Recognise water as a sacred gift to humanity
	• Recognise scarcity of water and legitimacy of common claim over it by all others in the world
	• Cultivate habits—conserve water and avoid over consumption and wastage
	• Proper maintenance is one of the most effective water savers
	• Decentralised toilet management systems like EcoSan and biogas plants
	• Eco-friendly materials for cleaning
	• Individuals to refrain from dirtying waterbodies by spitting, urinating, excreting, bathing, washing clothes and vessels and animals, and ritualistically throwing dead bodies into sacred rivers
Conservation	• Reverse osmosis purification
	• Drip irrigation
	• Solar
	• Gobar gas
	• Recycling and reuse
	• Water-saving technologies—water-saving toilets, water-saving showerheads, washing machines, spray wash for cars and intelligent tap systems
	• Recycling and reuse water
	• Conserve water and avoid wastage. Applicable to all individuals
Distribution	• Ancient scriptures indicate moral binding—since water is a life-giving element, and also since it is scarce, it needs to be made accessible to all. This means that no individual, household or even the community should have monopoly over water source. Even those who own waterbodies should realise since water is God-given, they are morally bound to share it with the needy persons. No individual can raise a dispute if the needy takes water for drinking and cooking purpose from out of the waterbody
	• Do not deny water to any thirsty (man or animal)
	• Equity to be ensured

Recent Initiatives

Scriptures have for sure shown great value in managing our water resources. There are initiatives taken on similar lines of scriptures but with a modern twist.

Table 2 Learnings for the community

Community: active role of NGOs/religious institutions and corporate social responsibilities	
Augmentation	• Engage voluntary labour to identify hillocks, forests and natural slopes by which rainwater drains out and build simple barrages to store rainwater to enhance groundwater table
	• Community must reserve land pockets (applicable to rural contexts) for a miniature forest which should remain untouched by people to ensure ecological balance
	• Explore possible options (like low-cost technologies and treatment plants) owned and run commonly by neighbourhood communities
Storage	• Protecting lakes as it ensures storage and groundwater recharge
	• Lakes have to be maintained by the community locally with free labour by the community (shramadhan was in practice)
	• Similarly maintaining village irrigation tanks by the community periodically through de-siltation
	• Identify places of water springs and groundwater sources and by contributing labour dig up open wells for storing water (applicable in rural contexts)
Protection	• As biblical scriptures indicate people were to assume a stewardship role with regard to nature, we should, as neighbourhood communities, take upon ourselves the responsibility of preserving, conserving and managing our neighbourhood sources of water
	• Non-governmental organisations and religious institutions can play an advocacy role for developing public policies in promoting local initiatives and contribute to develop water projects
	• Ensure that water sources are not polluted, and appropriate measures and adequate mechanisms are employed to keep our water sources such as tanks, ponds and lakes
	• Community should identify sources of water pollution—industries and habitual polluters—and take appropriate action to prevent water pollution and find ways of prevention through community enforcement systems
	• Creation and promotion of local communities and making them responsible and accountable for improve water governance
	• Water awareness community campaigns through programmes on saving, protecting and conserving water
Conservation	• Awareness modules for family members and children through visits to water resource management initiatives—tanks, sewage treatment plants, etc.
	• Engage in preservation and conservation of all neighbourhood water resources
	• Community should identify sources of water leakage and wastage and take necessary action to minimise such leakages and wastages
	• Community should educate on the need for water conservation
Distribution	• Community should identify cases where water is not shared with others (at least for drinking/cooking purpose) and take appropriate measures to make water accessible to them
	• For irrigating lands the community should ensure water access to reach tail-enders
	• Common wells have to be maintained in every area for distribution of water among the local citizens
	• Rationing of water should be done

Table 3 Learnings for the state

State—long-term—reuse, equity, tariff design, conservation options and incentives	
Augmentation	• State to have control over all water resources—ground and surface
	• Emphasise on revival of traditional systems, moral binding and the sprit behind it
	• Emphasise stakeholder participation at all levels and decentralisation in management—corporate, religious leaders, youth and women
	• Promote ancient Indian irrigation architecture, i.e decentralised systems
	• Establishment of sustainable water infrastructure
	• Tariff redesign for equity, slab rates and reused water
	• Tariff designs to ensure availability, minimum requirement for family/household at reasonable rates and on floor space/variety of uses (gardening, car washing, swimming pool)
	• Institutionalise reuse and recycle mechanisms and incentivise with rewards
	• Mandatory rainwater harvesting, recharge using treated waste water
	• Promote research in recycling technologies for community management
	• Ensure equitable distribution of water among regions/states
	• Regulations on industries to economise on usage and promote reuse
	• Ensure control of immoral and illegal practices
	• Maintain adequate forest cover to retain ecological balance
	• Preserving existing lakes and identification of areas for building lakes to store rainwater and avoid flooding
Storage	• Ensure economic, anthropogenic and geological feasibility in building large dams
	• Provide storage facilities like open wells, irrigation tanks, dams in rural areas and cisterns and storage tanks and lakes in urban areas for recharge and usage
Protection	• Divinity and morality to be strengthened towards natural resource management
	• Improving education, awareness and communication mechanisms for behavioural change
	• Establish effective water protection regimes for water protection
	• Water should not be allowed to be sold, traded, commoditised or exported for commercial gains
	• Water resource distribution to transnational corporations solely guided by profit maximisation to be avoided
	• Waterbodies under gram panchayat/town panchayath to entrust to village committees/civic bodies for maintenance/monitoring with participation
	• Stringent legislative measures for prevention of water pollution and appropriate institutional mechanisms to implementation of legislations
	• Maintaining lakes in every locality and treating waste waters at source
	• Promotion of gobar gas plants—domestic and community to save water
	• Sustainability through employment of local resources and local labour
	• Wastewater soaking pits
	• Increasing dependence on renewable energy—solar/wind
	• Innovative awareness programmes on behavioural change emphasising implications

(continued)

Table 3 (continued)

State—long-term—reuse, equity, tariff design, conservation options and incentives	
Conservation	• Conservation of water at the community level by the state by rationing water (particularly in the urban areas) where water is conveyed from distant sources
	• Regulate unaccountability and non-revenue water
	• Planting trees—converting urban parks to miniature forests—Nakshatravana, Nagavana
	• Awareness creation
	• Stringent regulations
Distribution	• Ensure social justice by ensuring equitable distribution
	• Subsidies for the poor and vulnerable and cross-subsidising others
	• Access to waterbodies to all, irrespective of caste
	• Conflicts and disputes to be settled by covenants
	• Better monitoring to check all malpractices/irregularities in distribution

Environmental movements in rural contexts and Green urbanism in cities have aided positive interventions. However, with competing water demands, there is a need for newer interventions to suit current contexts, for instance, improving tariff designs. In this context, everything may not be in line with the ancient scriptures where water was not charged. Water is a priced commodity in the given context, and hence with improved tariff designs, people will use water more judiciously. Documentation of traditional practices of water harvesting in Indian states by Centre for Science and Environment 'Dying Wisdom (Agarwal and Narain 1997)' is an interesting documentation. Such documentations would make people think and revive initiatives that suit our local contexts.

Similarly, given the magnitude of the challenges, advanced technologies of communication have made it easier for promotion of better water management. But, they have been brief, miniscule and one-time efforts cutting across state, institutions, communities and individuals. However, one need to acknowledge that there is for sure efforts to influence responsibility and improved participation. Under the Swachh Bharat Abhiyan initiative, Prime Minister Narendra Modi invited several public figures and its updates wherein celebrities, politicians and religious leaders have taken up the challenge to create and to promote clean India and sanitation for all. Several of the celebrities who have taken up the challenge update the various initiatives taken up so far and provide comments and suggestions. This has acted as an important platform to promote the programme. These campaigners of the programme also update the initiatives taken through other social networking sites like Facebook, Twitter, etc. that is more accessible to a larger section of society. Several independent app developers showed interest to develop *Swachh Bharat Apps* using mobile technologies. An article was published by Times of India on how 'Desi' companies beat Facebook in 'Swachh' app race. *Swachh Bharat Short Film* by *Ph.D.* students from NIT Rourkela highlighted the message that Swachh Bharat should not be a one day event but a part of our life to achieve the goal.

Promoting cleanliness across schools in the country under National School Sanitation Initiatives involving Bollywood actor Aamir Khan is promising. It is a joint programme of the Ministry of Urban Development (MoUD), Ministry of Human Resource Development (MHRD), Central Board for Secondary Education (CBSE) and the GTZ (Germany). On similar lines, improving children's lives under Total Sanitation Campaign in India, Pakistan, Sri Lanka, Bangladesh and Nepal by involving Indian cricket legend Sachin Tendulkar, UNICEF brand ambassador, is interesting.

There are green initiatives wherein media programmes in the recent past have been promoting messages on water conservation, for instance, programmes on celebrating 'waterless Holi' and 'Chemical-Free Holi Colours' supporting it with facts of water shortage and pollution. Besides this, we have star icons promoting for environmental concerns; the recent popular programme 'Satyameva Jayate' by Aamir Khan was popularly watched across TV channels. Vidya Balan the popular actress promoting the cause of sanitation is another interesting initiative taken up by the government under the Nirmal Bharat Abhiyan.

Interesting role of religious institutions can be seen all over the world, for instance, the cleaning of the Kumudvathi river, an initiative by the Art of Living Ashram of Sri Ravishankar Guruji, is an important contribution to reviving the river and promoting community responsibility and participation as it involved thousands of volunteers across the basin. Similarly, we have the Ecumenical Water Networks that promote awareness and training programmes, churches of South India and the Green Church as it has a 7-year plan for promoting ecological concerns. At Abu Dhabi, a mosque worked towards water saving through reuse during the daily ritual 'wudhu' which won a $500 as prize. There are several such instances of religious institutions working all over the globe to protect water resources which needs to be documented and popularised further for upscaling and promotion.

Green Temples are a promotion of the EXNORA—an NGO that promotes waste management in temples and sanitation complexes and protects temple tanks at various temples of Chennai. Technologies to compost temple waste and promoting reuse and sale are another interesting initiatives at various temples across India. At Delhi, 20,000 kg of flower waste polluted the river Yamuna, which has been curtailed with the temple machines converting temple waste into holy compost. In Uttar Pradesh, Gomti river is polluted less with the Pollution Control Board initiating temple waste management and reusing it in public parks as compost rather than dumping it in waterbodies.

Pollution Control Board has also been taking a massive awareness campaign in promotion of eco-friendly Ganeshas. This would help in protecting waterbodies to a large extent. Chemical dyes are polluting the waterbodies beyond rejuvenation. It would be useful to support the awareness campaigns with facts of contamination and implications for more effective control of using chemically coated painted Ganeshas for worship.

Various organisations like Infosys, Biocon, Arghyam, Parinaam, Ugly Truth and Pee project promote water and sanitation projects as CSR initiatives to protect dignity of the urban poor through construction of toilets and improve aesthetics of the city. Arghyam is working on a communication strategy for sanitation, in collaboration with the Government of Karnataka. This strategy aims to improve the quality of information, education and communication (IEC) under the Nirmal Bharat Abhiyan (NBA). Two communication agencies have been brought on board. 'Centre of Gravity' is to look into generating demand for toilets, while 'Final Mile' is to explore behavioural issues that prevent people from using toilets. Centre of Gravity would focus on developing a comprehensive communication campaign that can be executed by districts using the IEC budgetary allocation within the NBA. Final Mile would use a unique combination of cognitive neuroscience and behavioural economics to develop behavioural 'nudges', to get people to change their behaviours and use toilets.

Recent examples discussed so far show signs of hope as they are innovative initiatives supported by technology rooted in the combination of spirit to bring about the needed change. In addition, to make good water governance a reality, religion through scriptures helps us look back into the practices adopted, wherein moral-based approaches were more effective for reviving improved natural resource management. Good governance may be further strengthened by scaling up and drawing lessons from the roots of religion and influencing culture to attain the goals of sustainable development.

References

Agarwal A, Narain S (1997) Dying wisdom: rise, fall and potential of India's traditional water harvesting systems, vol 4. State of India's environment; a citizens' report. Centre for Science and Environment, New Delhi, p 404: fig., ill., photogr., tab

Appadurai, Breckenbridge (with Carol A. Breckenridge) (1990) Public culture in late 20th-century India. Items 44(4):77–80

Corcoran E, Nellemann C, Baker E, Bos R, Osborn D, Savelli H (eds) (2010). Sick water?: The central role of wastewater management in sustainable. A rapid response assessment. United Nations Environment Programme, UN-HABITAT, GRID-Arendal. ISBN:978-82-7701-075-5, Printed by Birkel and Trykkeri AS, Norway

Gleick PH (1998) Water in crisis: paths to sustainable water use. Ecol Appl 8(3):571–579

Guhan S (2007) The Cauvery Water Dispute: Towards Conciliation, A Frontline Publication. India's National Magazine, from the publishers of The Hindu, vol 24, Issue 03, 10–23 Feb 2007

Hall S (1997) The local and the global: globalization and ethnicity. In: King AD (ed) Culture, globalization and the world-system: contemporary conditions for the representation of identity. University of Minnesota Press, Minneapolis, MN.

IUCN (2000) Red list of Threatened Species, compiled by Craig Hilton Taylor, The IUCN World Conservation Union 2000.

Mahapatra R, Sakhuja N, Das S, Singh S (2008) The National Rural Employment Guarantee Act (NREGA) Opportunities and Challenges. CSE Policy Paper

Mehta VK, Goswami R, Kemp-Benedict E, Muddu S, Malghan D (2014) Metabolic urbanism and environmental justice: the water conundrum in Bangalore, India. Environ Justice 7(5):130–137. doi:10.1089/env.2014.0021

Martin D (1991) The cultural dimension of governance. In: Proceedings of the world bank annual conference on development economics. The World Bank, Washington, DC.

Narasimhan TN (2008) A note of India's Water Budget and Evapotranspiration. J Earth Syst Sci 117:237–240

Pringle CM, Freeman MC, Freeman BJ (2000) Landscape–scale effects of hydrological alterations on riverine macrobiota in the New World: tropical—temperate comparisons. BioScience September 2000:50(9)

Tomlinson J (1991) Cultural imperialism: a critical introduction. Continuum, London, New York, ISBN 0-86187-746-2

UNICEF, FAO, SaciWATERs (2013) Water in India: situation and prospects

Prakash A, Sharma M, Chourey J (2013) Water in India: situation and prospects. UNICEF, FAO and Saci Waters

Websites

Central Pollution Control Board. http://cpcb.nic.in/

Constitution of India. http://lawmin.nic.in/olwing/coi/coi-english/coi-4March2016.pdf

Inter-State Water Disputes Act, 1956Act 262(2). http://wrmin.nic.in/forms/list.aspx?lid=366

Inter-State Water Disputes Act, 1956. http://wrmin.nic.in/forms/list.aspx?lid=366

IUCN (2000) http://www.issg.org/pdf/guidelines_iucn.pdf

Karnataka Groundwater Regulation Act in 2012. http://www.cseindia.org/userfiles/KarnatakaGWact.pdf

MoWR (2006) http://wrmin.nic.in/

MNREGS. http://www.nrega.nic.in/netnrega/home.aspx

National Water Policy (2012) http://wrmin.nic.in/writereaddata/NationalWaterPolicy/NWP2012Eng6495132651.pdf

National Environmental Policy (2006) http://www.moef.gov.in/sites/default/files/introduction-nep2006e.pdf

UN. http://www.unfoundation.org/

World Water Council (2003) http://www.worldwatercouncil.org/index.php?id=192

Index

K.V. Raju, S. Manasi (eds.), *Water and Scriptures*,
DOI 10.1007/978-3-319-50562-6

MIX
Papier aus verantwortungsvollen Quellen
Paper from responsible sources
FSC® C105338

If you have any concerns about our products,
you can contact us on
ProductSafety@springernature.com

In case Publisher is established outside the EU,
the EU authorized representative is:
Springer Nature Customer Service Center GmbH
Europaplatz 3, 69115 Heidelberg, Germany

Printed by Libri Plureos GmbH
in Hamburg, Germany